THE USBORNE SCIENCE ENCYCLOPEDIA

Annabel Craig and Cliff Rosney

Designed by Steve Page and Russell Punter

Illustrated by Chris Lyon, John Shackell and Ian Jackson

Additional illustrations by
Peter Bull, Russell Punter, Robert Walster,
Steve Page, Martin Newton and Guy Smith

Contents

About this book

Scientists study the world around us. They look for explanations for everyday things, such as where lightning comes from and why rivers flow downhill. They make new discoveries and inventions, such as electricity, cars and computers, that change the way people live. This book answers many questions about the world around you and explains the science in everyday life. It is divided into eight sections. Each section is shown by a different color.

Counting and measuring

Heat and energy

Forces and machines

Light and color

Sound and hearing

Atoms and molecules

Electricity and technology

Lists and tables

On some pages there are quizzes. You can look up the answers on page 128.

You can tell which section you are reading by looking at the colored band along the top of the page.

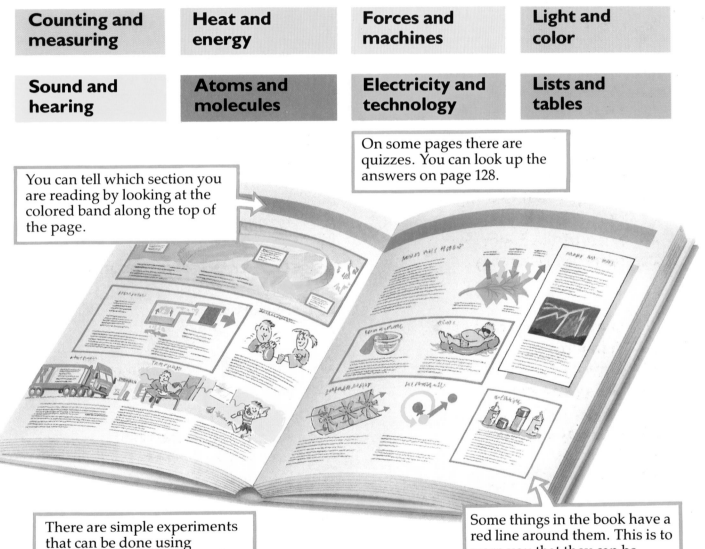

There are simple experiments that can be done using everyday things.

Some things in the book have a red line around them. This is to warn you that they can be dangerous.

Words are written in bold type where they are first explained. There is also a list of words with their explanations in the glossary on pages 116-119. There is an index on pages 120-127 to help you find things in the book. Some words have an asterisk after them, like this: gravity*. The footnote at the bottom of the page tells you where these words are explained in the book. Measurements are in metric units; the tables on pages 114-115 will help you to convert these to Imperial units. You can find out about these systems on page 7.

Footnotes are written at the bottom of the page.

Counting and numbers

People use numbers and counting so often in everyday life that it is difficult to imagine that they both had to be invented.

People had a rough idea of quantity before they invented numbers. They could tell that there were more animals in one herd than in another, but they could not count how many more. They could think in numbers of one, two, and perhaps three. They probably thought of more than three as just 'many'.

Keeping a tally

The first way people found to record an amount was to make a mark, like a scratch on a stick, for each item they were counting.

This is called keeping a **tally**. The Incas of Peru kept tallies of their animals and harvests by tying knots in cords. They called these cords **quipus**.

You probably also use tally marks sometimes. For instance, you may keep the score in a game by making a mark for each point a player makes.

Inventing numbers

After tally marks, people invented symbols, called **numerals**, to stand for various amounts. Different civilizations invented their own numerals.

Greek	A	B	Γ	Δ	E	F	Z	H	Θ	I
Roman	I	II	III	IV	V	VI	VII	VIII	IX	X
Hindu	٤	٤	٤	٤	٧	٤	೮	۲	٤	٤°
Medieval Arabic	1	2	3	٤	4	6	٨	8	9	10
Arabic numerals *	1	2	3	4	5	6	7	8	9	10
Binary *	1	10	11	100	101	110	111	1000	1001	1010

Roman numbers

Ancient Roman numerals are a mixture of tally marks and letters of the alphabet.

If the numeral on the right is smaller than, or equal to, the left one, you add them up.

If the numeral on the left is smaller, you subtract it from the one on the right.

Roman numerals were used in Europe for more than 1,500 years. Where can you see them today? (Answer on page 128.)

DID YOU KNOW?

The earliest known written numbers are about 5,000 years old and were found in the ancient city of Sumer (Iraq). They were scratched on wet clay tablets and then dried.

Changing numbers

The symbols we use to write numbers were invented about 1,500 years ago in India by Hindu mathematicians.

Arabs learned the numerals from them about 1,200 years ago.

Arab traders, 900 years ago, brought them to Europe. So they are often called **Arabic numerals**.

Arabic numbers are much shorter and simpler to write than Roman ones, because the value of each numeral changes, depending on its position. In Roman numerals, 2987 would be written as MMCMLXXXVII.

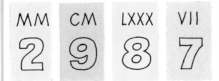

Arabic numerals have a symbol for zero. This makes it possible to show the difference between 2, 20, 200.

Different bases

We count in batches of 10, probably because we have ten fingers. This is called **base 10**, or the **decimal** system.

The Sumerians, 5,000 years ago, used base 60. It is the lowest number that can be divided equally by 2,3,4,5 and 6, so it is good for sharing things out.

Base 60 is still used today for measuring time. A minute has 60 seconds, and an hour has 60 minutes.

The binary system

Computers* and calculators use base 2, called the **binary system**, because they only use two symbols, 1 and 0.

Measuring things

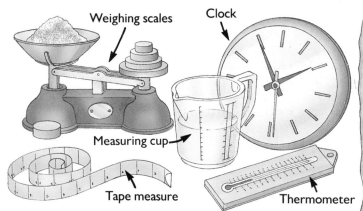

Weighing scales · Clock · Measuring cup · Tape measure · Thermometer

What time is it? How tall are you? How much do you weigh? What temperature is it outside? How far away are your nearest shops? You measure things every day. Measuring instruments help you measure things precisely.

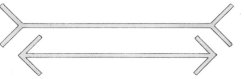

You cannot always believe what you think you see. Look at the two blue lines above: the top one looks longer than the bottom one. But if you measure them with a ruler, you will find that they are both the same length.

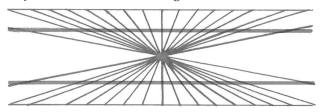

Move the book around and look at the two red lines above from different angles. The lines look as though they bend a bit in the middle. But they are, in fact, straight and parallel.

You cannot always rely on what you feel either. You may feel that it is cold outside, but the same air may feel warm to someone else. A thermometer measures the temperature exactly.

Body measurements

When you measure something, you are really comparing it to a fixed quantity, like a metre. This is called a **unit of measurement**. The first units of measurement were based on the body. The Ancient Egyptians used units of measurement called **cubits**, **palms** and **digits**.

Ancient Egyptian measurements

One digit

One palm

One palm = four digits

One cubit = seven palms

One cubit = elbow to tip of middle finger.

Roman measurements

The Romans used the length of a foot to measure distance. To measure smaller lengths, they divided one **foot** into 12 thumb widths. They called each thumb width an **uncia**.

'Inch' comes from the Latin word *uncia*.

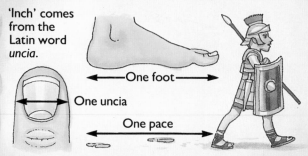

One foot

One uncia

One pace

They measured longer distances in **paces**, counting each pace as two steps. They called 1,000 paces a **mile**. The word 'mile' comes from a Latin word, *mille*, meaning one thousand.

Yards

One yard

Cloth traders invented a unit of measurement called the **yard**. Every yard was a length of fabric stretched between chin and fingertips.

Imperial units

Any unit of measurement can be used to measure things, as long as other people can use the same unit. The problem with measurements based on the body is that they vary, depending on people's sizes. About 900 years ago, King

Henry I of England made a law to make all yards the same length, this was the length between *his* chin and fingertips. Later, more laws fixed other measurements. They became known as **Imperial units** and are still used in some countries today.

How heavy you are is measured in stones, pounds and ounces.

Distance is measured in miles, yards, feet and inches.

Volume is measured in gallons, pints and fluid ounces.

The metric system

The first unit of measurement that was not based on the body was a unit of length called the **metre**. The **metric system** of measurement is based on the metre.

North Pole

Paris

Equator

A platinum bar was made exactly one metre long. Copies were made of it so that a record of a metre could be kept in different places.

The metre was invented about 200 years ago in France. It was calculated by dividing the distance between the North Pole and the Equator, through Paris, by 10 million.

Today, the metre is fixed by measuring how far light travels in a set time.

One metre

Most countries today use the metric system. Buying and selling between countries is much easier if everyone uses the same system.

How tall are you?

Lie on the floor and get a few people of different sizes to measure your height. First ask them to use Egyptian cubits, palms and digits, and then Roman feet and *uncia*. How different are their answers and why? (Answer on page 128.)

Time

Thousands of years ago, people did not need to measure time in any detail. They only needed to count days and nights and observe the seasons to know when to plant their crops.

Today, time is measured very precisely in units of hours, minutes and seconds. You can see this on train and bus timetables. They show the departure and arrival times to the minute.

The Egyptian year

As long as 5,000 years ago, the Ancient Egyptians divided their calendar into 365 days. They noticed that every 365 days a star called Sirius appeared in the sky just before sunrise.

They knew that, at about the same time the star appeared, the River Nile would flood. After the floods, the farmers were able to plow their fields and plant their crops.

Measuring time

1. Egyptian shadow clocks are the earliest known clocks and were used 4,000 years ago. The time was read from the shadow on the scale.

2. Water clocks were used by the Egyptians on cloudy days or at night. As water trickled out of a stone pot, the time was shown by the water level.

3. Candle clocks were invented about 1,000 years ago. As the candle burned down, it showed how many hours had passed.

4. Pendulum clocks were the first clocks able to measure seconds. Galileo invented the pendulum, but the first pendulum clock was made by Christiaan Huygens in 1667.

5. Quartz crystal clocks were first made in 1929. The first quartz crystal wristwatches were made in 1969. They keep time very accurately.

6. Atomic clocks are used by scientists to measure time very precisely. They are accurate to about one second every 300,000 years. The first was built in 1948.

Why are there days and nights?

The Earth spins on an imaginary line called the **axis**. The side facing the sun has daylight, while the other side is dark. It takes the Earth 24 hours to complete each spin.

Leap years

The Earth takes 365¼ days to orbit the Sun, but a calendar year has only 365 days. So every four years, an extra day is added to February. These years have 366 days and are called **leap years**.

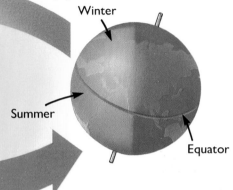

Changing seasons

The Earth leans to one side, so one half is tilted nearer to the Sun. That half has summer while the other half has winter.

As the Earth orbits the Sun, a different part of the Earth is gradually tilted nearer the Sun. This makes the seasons change.

Places on the Equator do not have summers or winters because they always stay the same distance from the Sun.

Time zones

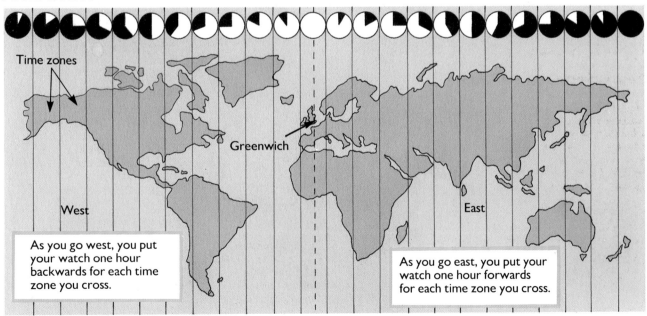

As you go west, you put your watch one hour backwards for each time zone you cross.

As you go east, you put your watch one hour forwards for each time zone you cross.

The world is divided into 24 **time zones**. The time is counted from Greenwich, London. The time zone east of Greenwich is an hour ahead, the one to the west is an hour behind Greenwich.

What is energy?

All around you things are happening. The wind may be blowing, cars may be driving by, and people talking and moving about. As you read this, your eyes are moving and blood is flowing around your body.

In this picture, you can see many different types of energy making lots of different things happen. All these things are happening because of **energy**. Energy is what makes things happen all over the Earth and the Universe.

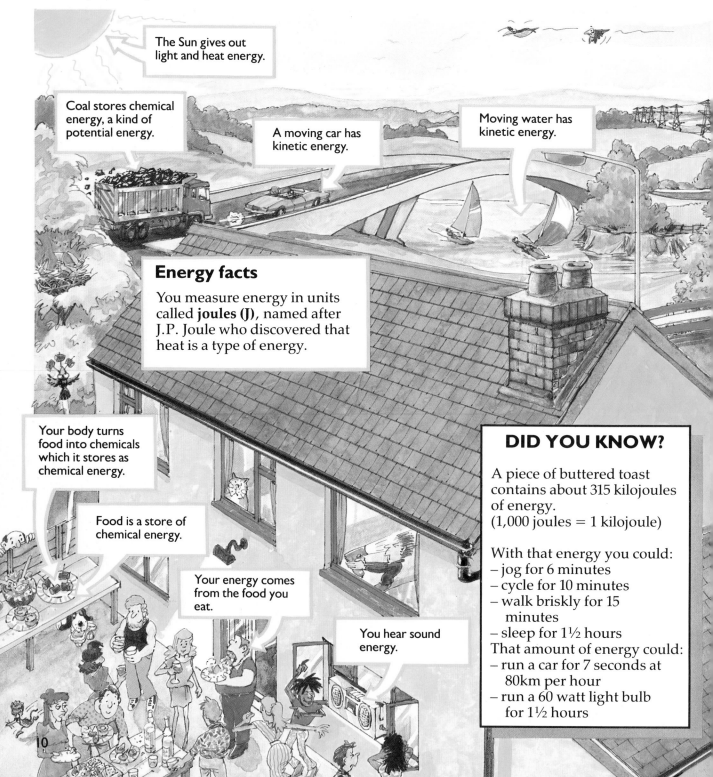

The Sun gives out light and heat energy.

Coal stores chemical energy, a kind of potential energy.

A moving car has kinetic energy.

Moving water has kinetic energy.

Energy facts

You measure energy in units called **joules (J)**, named after J.P. Joule who discovered that heat is a type of energy.

Your body turns food into chemicals which it stores as chemical energy.

Food is a store of chemical energy.

Your energy comes from the food you eat.

You hear sound energy.

DID YOU KNOW?

A piece of buttered toast contains about 315 kilojoules of energy.
(1,000 joules = 1 kilojoule)

With that energy you could:
– jog for 6 minutes
– cycle for 10 minutes
– walk briskly for 15 minutes
– sleep for 1½ hours
That amount of energy could:
– run a car for 7 seconds at 80km per hour
– run a 60 watt light bulb for 1½ hours

Energy is needed to make cars go, to heat and light your home and to keep your body working. The different types of energy can be divided into two groups, depending on whether the energy is moving or stored.

Energy that is moving is also called **kinetic energy**. Energy that is stored is also called **potential energy**. You can find out more about different types of energy and how they are used in the next few pages.

Electric lights give out light energy.

Electrical energy flows through wires to homes and factories.

Wind, or moving air, has kinetic energy.

Oil, coal, wood, gas and all other fuels store chemical energy.

Energy quiz

Not all the examples of energy in this picture are labelled. See how many others you can find. (Answers on page 128.)

Anything that could fall has potential energy. The higher it is, the further it could fall, so the more potential energy it has.

A stretched elastic has potential energy.

Batteries in a flashlight store chemical energy.

Fire gives out heat and light energy.

Anything falling has kinetic energy.

Energy changes

All the different sorts of energy around you can be changed into other sorts of energy.

In fact, energy cannot be made or destroyed, it can only be changed into another sort of energy.

If you eat too much, your body stores the extra food as fat.

Your energy comes from the food you eat. Your body changes chemical energy inside food into a different sort of chemical energy, and stores it.

An electric clock works because chemical energy inside its batteries* is changed to electrical energy*. When the alarm rings, electrical energy is changed to sound energy*.

The sound of your voice is changed into electrical energy...

...and the electrical energy is changed back to sound energy.

Telephones change sound energy into electrical energy, and electrical energy back into sound energy.

When you move, your body changes chemical energy from the food you eat into moving, or **kinetic**, energy*.

Cars need chemical energy stored in fuel to move. The engine* changes chemical energy into kinetic energy.

Power stations change the chemical energy of fuel or the kinetic energy of moving water into electrical energy.

Nuclear energy* can be changed into electrical energy. Solar panels change the Sun's heat energy into electrical energy.

Electrical energy is changed into light energy* by light bulbs, and into heat energy by heaters.

*Batteries, 95; Electricity, 92; Engines, 45; Kinetic energy, 11; Light energy, 50; Nuclear energy, 77; Sound energy, 64.

Electric ovens, toasters and irons change electrical energy into heat energy*. An electric whisk changes electrical

energy into kinetic energy. Televisions* change electrical energy into light and sound energy.

When a firework explodes, its chemical energy changes into light energy, sound energy and heat energy.

The potential, or stored, energy* of anything that could fall changes into kinetic energy as it begins to fall.

Find the energy changes

What sorts of energy are being changed into other sorts of energy in this picture? (Answers on page 128.)

DID YOU KNOW?

When you run, only about 25% of the chemical energy in your muscles changes into kinetic energy. The rest changes into heat energy.

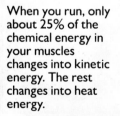

People need to change energy from one sort to another to do many things. But not all of the energy changes into the sort of energy they want.

How much energy you get out compared to how much you put in is called **efficiency**. In most cars only about a quarter, or 25%, of the chemical energy

of gasoline changes into kinetic energy. The rest is lost as heat energy and sound energy. Cars are only about 25% efficient.

Heat and temperature

Heat is a form of energy. You use it for lots of everyday things like keeping warm, heating water and cooking food.

Heat moves

Heat energy does not stay still, it moves. It spreads out from hotter things to cooler things until both are the same temperature.

Leave a hot and cold drink out for a few hours. The hot drink cools down and the cold drink warms up until both reach room temperature.

Heat energy moves in three ways, by **conduction** (see below), by **convection*** and by **radiation***.

Conduction

Stir a hot drink with a metal spoon. The handle gets hot because heat travels through it. This is called **conduction**. Heat moves through solids by conduction. Through some solids, like metals, it moves very quickly. These are called good **conductors**. Other solids, like plastic, are poor conductors. These are called **insulators**.

Saucepans are made of metal so that they conduct heat to the food to cook it.

Pan handles are made of plastic or wood because they are insulators.

Why does metal feel cold?

When you touch metal, it feels cold. Because metal is a good conductor, the heat from your hand flows out into it. It is not the metal that is cold, it is your hand losing heat.

Convection, 16; Radiation, 18.

Air can keep you warm

Your clothes keep you warm because they stop you losing your body heat. This is because they trap air. Your body heat cannot get through the trapped air, because air is an insulator.

Snow is a good insulator, because it traps lots of air.

People lost in blizzards dig holes in the snow for warmth.

Walls have a gap filled with air for insulation.

Feather-filled jackets keep you warm because they trap a lot of air.

Thick winter clothes trap lots of air.

Birds fluff up their feathers in winter to trap more air.

Wool feels warm because it traps lots of air in its fibers.

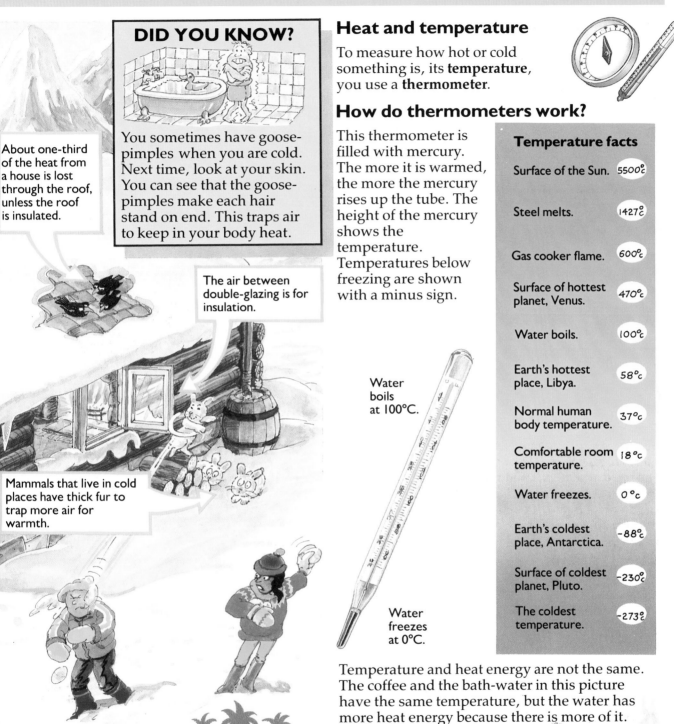

About one-third of the heat from a house is lost through the roof, unless the roof is insulated.

The air between double-glazing is for insulation.

Mammals that live in cold places have thick fur to trap more air for warmth.

Heat and temperature

To measure how hot or cold something is, its **temperature**, you use a **thermometer**.

How do thermometers work?

This thermometer is filled with mercury. The more it is warmed, the more the mercury rises up the tube. The height of the mercury shows the temperature. Temperatures below freezing are shown with a minus sign.

Water boils at 100°C.

Water freezes at 0°C.

Temperature facts

Surface of the Sun.	5500°c
Steel melts.	1427°c
Gas cooker flame.	600°c
Surface of hottest planet, Venus.	470°c
Water boils.	100°c
Earth's hottest place, Libya.	58°c
Normal human body temperature.	37°c
Comfortable room temperature.	18°c
Water freezes.	0°c
Earth's coldest place, Antarctica.	-88°c
Surface of coldest planet, Pluto.	-230°c
The coldest temperature.	-273°c

Temperature and heat energy are not the same. The coffee and the bath-water in this picture have the same temperature, but the water has more heat energy because there is more of it.

You measure temperature in units called **degrees Celsius (°C)** and heat energy in units called **Joules (J)**.

Air can keep you cool

As well as keeping you warm, air can keep you cool. In hot countries, people wear loose clothes so that air can circulate. This stops heat from the Sun being conducted to their bodies.

Heating air and water

Gases, like air, and liquids, like water, are usually bad conductors* of heat. This means that if they are trapped so that they cannot move, heat does not pass through them easily. But if a gas or liquid is free to move, it can carry heat energy with it. A heater is able to heat a whole room because the air in the room is free to move. When you turn a heater on, moving air carries heat energy from the heater to all parts of the room.

How does heat move around this room?

The heater warms up the air just around it. This heated air rises, because hot air is lighter than cold air.

As the warm air rises, cold air sinks down to take its place. This cold air is then warmed by the heater, and in turn rises.

Soon, air is moving around the room, carrying heat energy with it, until the temperature of the whole room is higher.

This moving air is called a **convection current**. The air in the room has been heated by **convection**.

How is water heated up?

Heat energy also moves through liquids by convection. When a pan of water is heated, the pan heats up first. This happens by conduction*. The hot pan then heats the water next to it.

The heated water rises, and cold water takes its place. This is because warm water is lighter than cold water. The water starts to move, setting up a convection current. All the water heats up eventually.

Watch heat move

Hold a piece of tissue paper over a heater and watch the convection current make it flutter.

Look at something that is hot. The air above it shimmers. This is the hot, lighter air rising through the colder air.

On very hot days, road surfaces sometimes become so hot that you can see the air shimmering above them.

*Conduction, 14; Conductors, 14.

Why does smoke rise?

Smoke rises from a bonfire by convection. You can often see pieces of ash float up with the smoke.

Volcanic ash

When volcanoes erupt, they set up very strong convection currents which send ash and dust high up in the sky.

In 1980, Mount St Helen's, USA, erupted. It sent a blast of ash 9km above the Earth, blocking out heat and light from the Sun.

Wind* is simply moving air. It is made by convection currents over the surface of the Earth. Land heats up more quickly than sea. On a hot, sunny day, warm air above the land rises and cold air blows in from the sea, taking its place. Land also cools down faster than sea, so at night the opposite happens. Warm air rises above the sea, and cold air blows out from the land.

Keeping up with convection

A glider is towed into the air by a powered plane, which then lets it go.

Glider pilots often find thermals by watching birds flying.

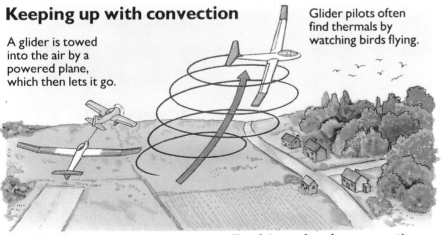

A glider has no engine to power it, but it can fly for long distances and even climb higher. This is because of the warm air that rises from the Earth's surface by convection. These convection currents are called **thermals**. A glider is able to fly for as long as the pilot can find thermals.

DID YOU KNOW?

Swifts fly non-stop for their first two or three years, until they are old enough to breed. They eat and drink while flying. At night, they rest on thermals high in the sky. (Not, of course, as shown in the picture, but in a gliding position.)

Heat rays

When you stand in sunshine, the sunlight feels warm, because you are taking in heat energy from the Sun. This heat energy passes through 150 million kilometres of Space to reach Earth. Heat energy cannot reach Earth by conduction* or convection* because Space is empty. The heat travels to Earth in invisible straight lines, called heat rays, that spread out, or radiate, from the Sun. When heat moves this way, it is called heat radiation.

Heat rays from the Sun

The temperature at the centre of the Sun is 16 million °C.

Most of the Sun's heat is absorbed by the **atmosphere*** which surrounds the Earth.

The Earth is heated by the Sun.

Less than one millionth of the Sun's radiation reaches Earth.

Some heat from the Sun is **reflected** away from Earth.

Some of the Earth's heat radiates away.

The Earth takes in, or **absorbs**, some of the Sun's heat.

If some heat did not radiate away, the Earth would just get hotter and hotter. Clouds help keep the heat in, but they also block out some of the heat radiation from the Sun.

How does a broiler work?

A broiler cooks food by heat radiation. The food absorbs heat rays from the broiler.

The broiler stays hot after you have switched it off, until the heat has radiated away.

The heat moves downwards to the food. It cannot be moving by convection because convection carries heat upwards. It cannot be moving by conduction because air is a good insulator and does not conduct heat well.

Heat pictures

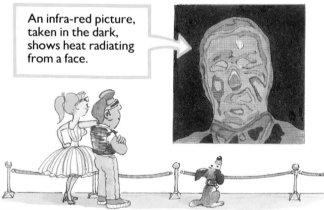

An infra-red picture, taken in the dark, shows heat radiating from a face.

Heat rays are also called **infra-red rays***. A photograph can be taken of heat with a special infra-red camera. Different colors show the amount of heat that radiates from things. Hot things radiate more heat than cold things.

Absorption and reflection

The more heat rays that something absorbs, the hotter it becomes. Things that reflect some of the radiation away will not become as hot. Some surfaces absorb more heat rays than others. Dull and dark surfaces absorb more heat rays than shiny and light surfaces, which reflect them away.

Heat reflected from surface

Heat absorbed by surface

Red tiles

Black things left in the Sun feel hotter than white things.

Shiny metal

White wall

Concrete

In hot countries, people paint their houses white to reflect heat away.

Soil

Shiny spacesuits

There is no atmosphere around the Moon to absorb the Sun's heat radiation. This means that the Sun feels much hotter there. To help keep them cool, astronauts wear shiny suits that reflect the heat away.

Soot and snow

Black things absorb much more heat radiation than white things. So snow will melt more quickly in sunshine if you put soot on it.

Weather satellites

Scientists who study weather are called **meteorologists**. They use infra-red photographs taken by satellites to help them make weather forecasts. There

Dark parts stand for warm places.

Bright parts stand for cool places.

Polar satellites move around, or orbit the Earth, passing over the North and South Poles on each orbit. They are able to

are two types of weather satellites, **geo-stationary** and **polar satellites**. Geo-stationary satellites stay fixed, about 35,000km above the Equator.

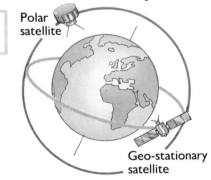

Polar satellite

Geo-stationary satellite

photograph the whole of the Earth's surface, because the Earth spins on its axis beneath them.

DID YOU KNOW?

Some burglar alarms work by detecting infra-red rays. The alarm goes off when it detects the heat radiating from a burglar's body.

Energy in living things

The living world of plants and animals stretches from the bottom of the deepest oceans to the top of the highest mountains. Every plant and animal in it needs energy to keep alive. Their energy comes from food. All that food depends on the Sun's energy.

Food chains and food webs

Green plants are able to change the Sun's light energy into chemical energy which they use as food. They are the only living things that can do this.

Some animals eat green plants, and they, in turn, are eaten by other animals. In this way, the Sun's energy passes from one living thing to another. This is called a **food chain**.

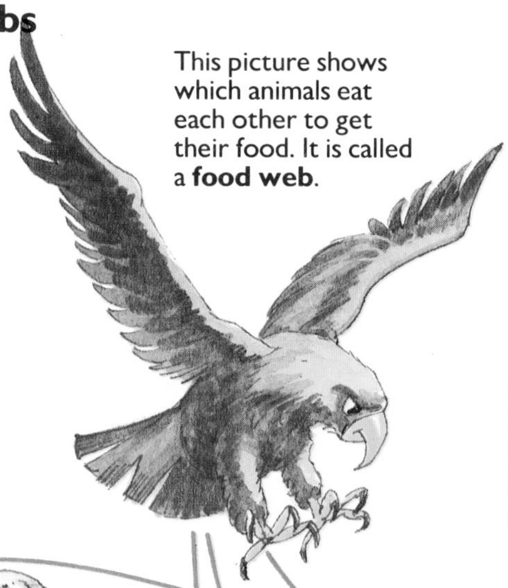

This picture shows which animals eat each other to get their food. It is called a **food web**.

DID YOU KNOW?

Nearly three-quarters of the Earth is covered by sea. Tiny plants called **plankton** live in the sea and make nearly three-quarters of the Earth's oxygen. The biggest is about 1mm across, the smallest is about fifty times smaller.

Rabbits eat lettuces.
Lettuce gets its energy from the Sun.
Foxes eat rabbits.

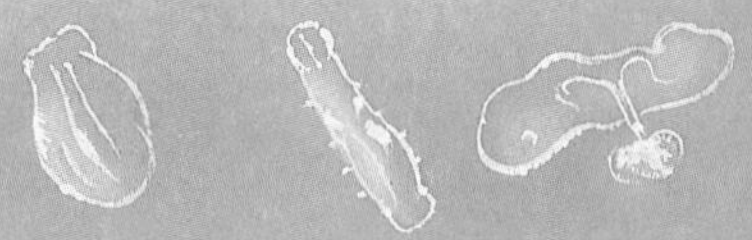

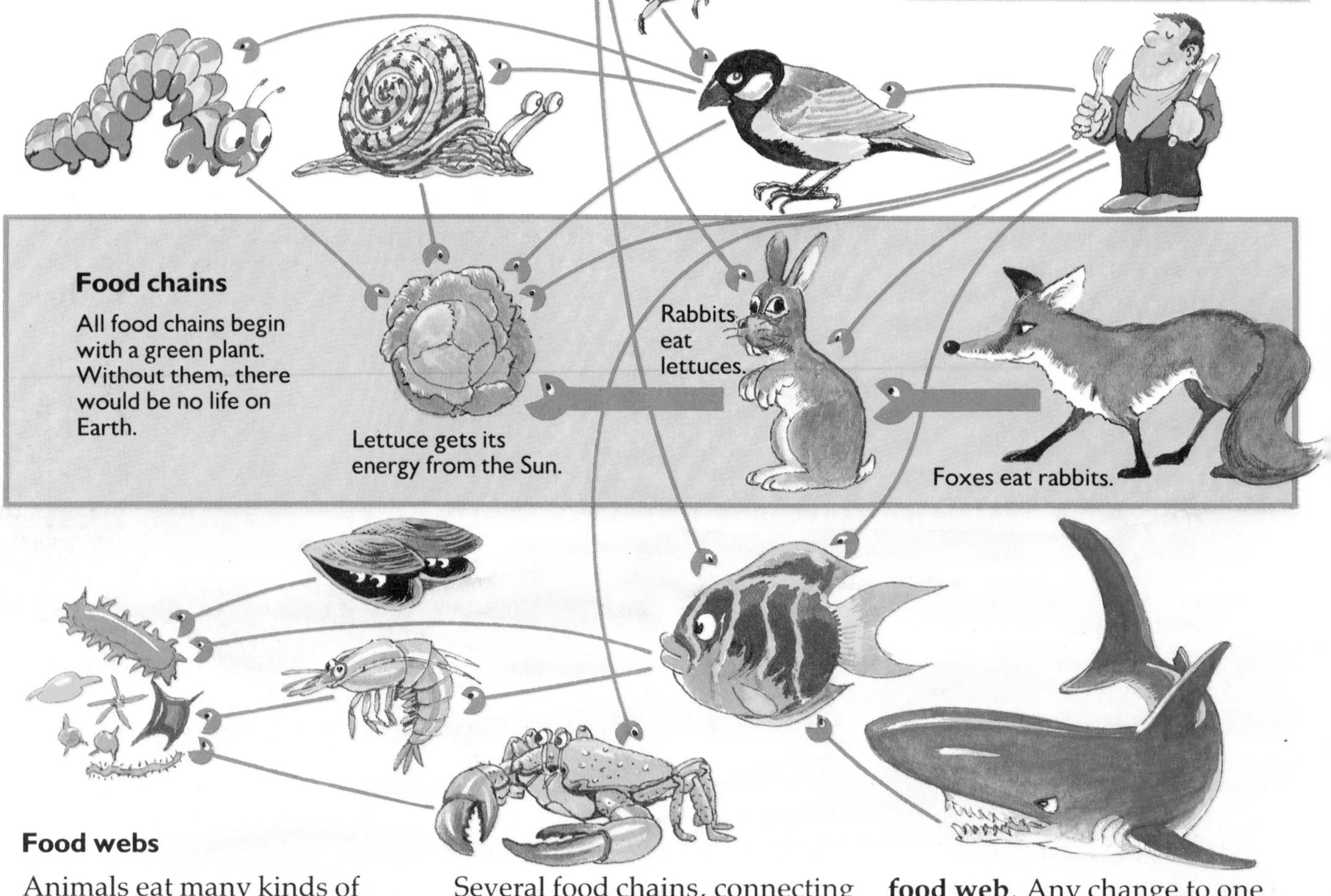

Food chains

All food chains begin with a green plant. Without them, there would be no life on Earth.

Food webs

Animals eat many kinds of food, so each animal belongs to many different food chains. Several food chains, connecting the lives of many different plants and animals, are called a **food web**. Any change to one part of a food web can change the lives of the other things in it.

How do plants make their food?

Green plants make their own food. They take in sunlight and a gas, **carbon dioxide**, from the air. Sunlight and carbon dioxide join up with water and a chemical, **chlorophyll**, inside the leaves. This makes the plant's food, a sugar called **glucose**. At the same time, the plant gives out oxygen through its leaves. This is called **photosynthesis**.

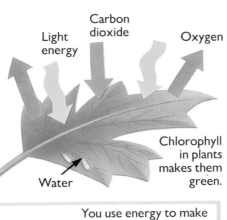

Light energy

Carbon dioxide

Oxygen

Chlorophyll in plants makes them green.

Water

Energy from plants to animals

When animals eat green plants, glucose from the plant joins with oxygen inside their bodies. This is how they get their energy. While this happens, carbon dioxide and water are also formed. This way of changing food back into energy is called **respiration**.

Why do you breathe?

You breathe in because your body needs the oxygen in air for respiration. This is how you get your energy. You breathe out to get rid of the carbon dioxide and water made during respiration. Breathe out on to a mirror. The moisture you see is the water made by respiration.

Why do you need food?

You use energy to make your muscles move and to keep your body warm.

Water   Protein

Minerals

Roughage

Carbohydrate

Vitamins

Fat

You need different foods to keep your body healthy. Your energy comes from foods that contain carbohydrates and fats.

Your body also needs **proteins** to grow and to repair itself, as well as **vitamins**, **minerals**, **roughage** and water.

Plants in the dark

At night, plants take in oxygen.

They give out water and carbon dioxide.

In daylight, plants make their food by photosynthesis. At night, when there is no light, they take in oxygen to get their energy by respiration.

Balancing gases in the air

Carbon dioxide

Oxygen

Green plants make more oxygen during the day than they use at night.

Oxygen and carbon dioxide are always being added to the air and taken from it by living things. Green plants make all the Earth's oxygen by photosynthesis during the day. People and animals need to breathe oxygen to stay alive, so without plants there could be no animal life on Earth.

Planet Earth

The story of the Earth

The Earth was formed about 4,500 million years ago. One theory is that the Earth started off as a huge, swirling cloud of dust and gases.

The cloud began to shrink and turned into a ball of hot, liquid rock.

When the surface cooled, it turned to a solid crust of rock that gave off clouds of steam and gases.

4,500 million years ago

570 million years ago

340 million years ago

Heavy rain poured down from the clouds. It flooded the Earth, forming the first seas.

280 million years ago

50 million years ago

The Earth's distance from the Sun makes it just the right temperature for life to exist.

Fossils are the remains of early plants or animals left in ancient rocks. Scientists are able to build up a picture of life millions of years ago by studying fossils.

Land is divided into seven **continents**. Over millions of years, they have slowly moved over the Earth's surface. This is called **continental drift**.

The Earth's surface is still changing today. Every year the Atlantic Ocean gets about 40mm wider. In a million years it will be 40km wider.

DID YOU KNOW?

It has taken millions of years for the seas to become salty. The water from rain and melting snow has gradually dissolved salt from rocks, and the salt has built up in the seas.

The planet Earth

The Earth is one of nine planets that orbit the Sun. This is called our **solar system**.

The Sun is a star, just like the stars you see at night. It looks much brighter because it is nearer.

Scientists believe that the Sun was formed about 5,000 million years ago, when a large cloud of gas began to shrink and heat up.

The Sun is 150 million km away. The nearest star after that is 40 million million km away.

Stars are gathered in groups, called **galaxies**. There are millions of stars in each galaxy and millions of galaxies in the Universe. Our solar system is in a galaxy called the **Milky Way**.

The changing Earth

The Earth's crust is made of separate pieces called **plates**, which float on the hot magma. The plates fit together like a giant jigsaw puzzle.

Where plates have been squeezed together over millions of years, they have folded over each other, forming mountains.

Most **earthquakes** happen near the edge of plates where there are cracks called **faults**. Earthquakes are caused when the plates move against each other.

There are a total of 46 moons in our solar system. The Earth has one moon orbiting it, but Jupiter has 14 moons.

What is the Earth made of?

The Earth is a huge ball of rock. There are three parts, the **crust**, **mantle** and **core**.

The crust under the sea is about 6km thick.

The top layer of the **mantle** is hot, liquid rock, called **magma**. The **crust** floats on it.

The deepest hole ever dug is 13km deep. Its temperature is 200°C.

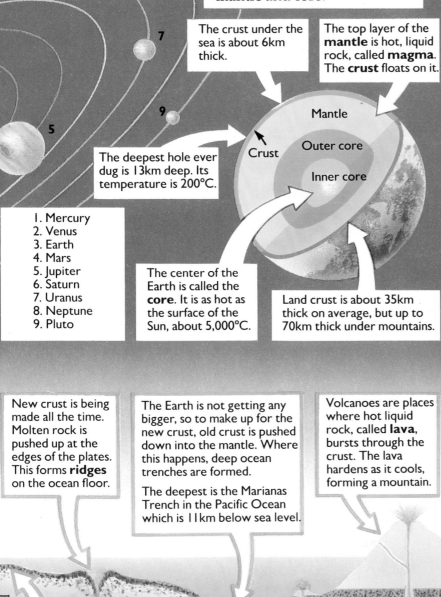

Mantle

Outer core

Inner core

Crust

1. Mercury
2. Venus
3. Earth
4. Mars
5. Jupiter
6. Saturn
7. Uranus
8. Neptune
9. Pluto

The center of the Earth is called the **core**. It is as hot as the surface of the Sun, about 5,000°C.

Land crust is about 35km thick on average, but up to 70km thick under mountains.

New crust is being made all the time. Molten rock is pushed up at the edges of the plates. This forms **ridges** on the ocean floor.

The Earth is not getting any bigger, so to make up for the new crust, old crust is pushed down into the mantle. Where this happens, deep ocean trenches are formed.

The deepest is the Marianas Trench in the Pacific Ocean which is 11km below sea level.

Volcanoes are places where hot liquid rock, called **lava**, bursts through the crust. The lava hardens as it cools, forming a mountain.

Near land, the oceans are about 200m deep. This area is called the **continental shelf**. Beyond that, the average depth of the oceans is 5,000m.

Most volcanoes are under the sea. They form near the edges of the plates which are mostly underwater.

The atmosphere

The Earth is surrounded by a layer of air about 10,000km thick, called the **atmosphere**. Air is a mixture of **gases**. The main gases are **nitrogen**, **oxygen**, **argon** and **carbon dioxide**.

The atmosphere is held around the Earth by gravity*. There is less air higher up and the atmosphere gradually blends into Space, where there is no air at all.

The **ionosphere** is about 450km thick. Radio waves* travel around the Earth by bouncing off it.

Jets fly in the lowest part of the **stratosphere**, which is about 45km thick. This is above the changing weather.

About 20km above the Earth, there is a thin layer of a gas, called **ozone**. It protects Earth from harmful ultraviolet rays* from the Sun.

The bottom layer of the atmosphere, the **troposphere**, is about 10km thick. This is where the weather* happens.

Ionosphere

Stratosphere

Ozone layer

Troposphere

A blanket around the Earth

The atmosphere acts as a layer of insulation between the Earth and the Sun. During the day, it protects the Earth from the burning heat of the Sun. At night, it acts like a blanket, keeping in the heat absorbed from the Sun during the day.

*Gravity, 32; Radio waves, 106; Ultraviolet rays, 104; Weather, 84.   23

Fuels of the Earth

A huge amount of energy is needed to run all the world's machines and industries. Most of it comes from three fuels: oil, coal and gas. These fuels are used to heat houses, to drive cars and to make electricity. Oil, coal and gas are called fossil fuels because they were formed from the remains of prehistoric plants and animals.

How old is a piece of coal?

About 300 million years ago, the Earth was covered in swampy forests full of giant plants. As the plants died, they were buried under mud.

The mud gradually hardened into rock. The rotting plants were squashed between heavy layers of rock and heated by the Earth. Over millions of years they changed into coal.

Coal mines

Coal is collected from pits or mines deep underground. Miners use explosives to blast the rock and cut the coal out with machines.

Fossil hunt

If you look carefully at lumps of coal, you may be able to find a fossil of a leaf that lived many millions of years ago.

Oil and gas

Oil and gas were formed over millions of years. They come from the remains of tiny animals that lived in the prehistoric seas. Gas formed as the animals rotted away.

Oil is reached by drilling holes in the ground. The oil may gush to the surface or it may have to be pumped up.

Nearly half the world's oil is found under the sea floor. It is reached from enormous oil rigs. They are among the largest structures ever built.

Rigs are also used to drill for gas, in the same way as oil. The gas is then piped to tanks on land.

Oil rig

Surplus gas being burned off

Helicopter landing pad

Crane

Rock

Gas

Drill

Oil

Finding fossil fuels

Coal, oil and gas are not always found at the same depths under the Earth's surface. This is because the Earth's crust* has changed over millions of years. Places that were land are now sea. Others that were sea are now land.

When the fuels run out

Fossil fuels supply three-quarters of the Earth's energy. They took millions of years to form, so they cannot be replaced when they run out.

The Earth's coal supply has been used for many hundreds of years. There is probably enough left to last for another thousand years.

People only began to use oil as a fuel after car engines* were invented, about 100 years ago. There may only be enough to last for another 60 years.

*Car engine, 45; Earth's crust, 22

Using fossil fuels

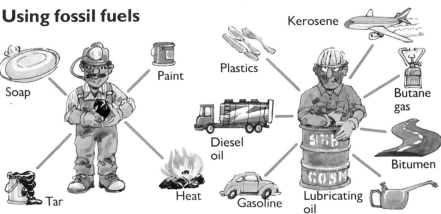

Coal is burned to make heat. But it can also be used to make other useful things. It is used to make soap, dyes, perfumes, paints, tar and many chemicals.

Oil from the ground is called **petroleum**, or **crude oil**. It is a mixture of useful chemicals. These are separated, or **refined**, in a place called a **refinery**.

Pollution

To get energy out of a fossil fuel, it has to be burned. The heat of the burning fuel can be used for warming something up, or it can be used to make an engine work.

When fossil fuels are burned, they dirty, or **pollute**, the air. They let off smoke and gases that are very harmful to people, plants and animals. This is called **pollution**.

When gasoline is burned in car engines, it makes a very poisonous gas called **carbon monoxide**. Tiny specks of soot from burning coal dirty the air.

Burning coal also makes a gas, called **sulphur dioxide**. It causes **acid rain**, which poisons trees and plants, and wears away metal and stone.

Nuclear energy

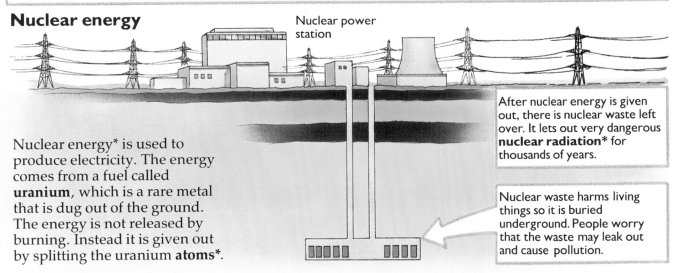

Nuclear power station

Nuclear energy* is used to produce electricity. The energy comes from a fuel called **uranium**, which is a rare metal that is dug out of the ground. The energy is not released by burning. Instead it is given out by splitting the uranium **atoms***.

After nuclear energy is given out, there is nuclear waste left over. It lets out very dangerous **nuclear radiation*** for thousands of years.

Nuclear waste harms living things so it is buried underground. People worry that the waste may leak out and cause pollution.

*Atoms, 76; Nuclear energy, 77; Nuclear radiation, 77.   25

Alternative energy

Fossil fuels cause harmful pollution and are running out. So people are finding new kinds of energy to produce electricity and run machines.

Energy that does not come from oil, coal, gas or nuclear power is called alternative energy. It mostly comes from water, sun and wind.

Water power

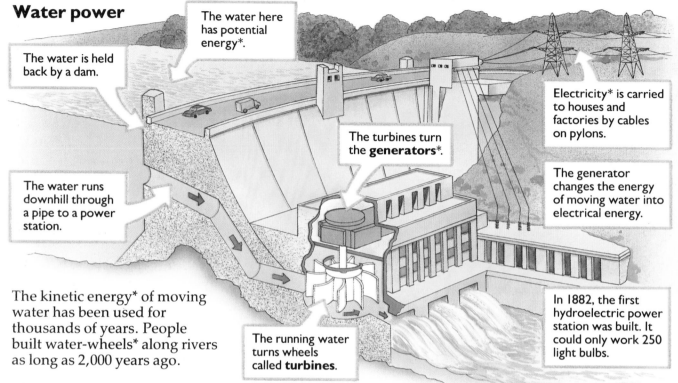

The water here has potential energy*.

The water is held back by a dam.

The water runs downhill through a pipe to a power station.

The turbines turn the **generators***.

Electricity* is carried to houses and factories by cables on pylons.

The generator changes the energy of moving water into electrical energy.

The running water turns wheels called **turbines**.

In 1882, the first hydroelectric power station was built. It could only work 250 light bulbs.

The kinetic energy* of moving water has been used for thousands of years. People built water-wheels* along rivers as long as 2,000 years ago.

The energy of moving water is now used to produce electricity in **hydroelectric power stations**. Hydroelectricity provides over six per cent of the energy used in the world today. Because the water comes from rain or melting ice, it never runs out. Only countries that have lots of water can produce electricity this way. Scandinavia, North America and the USSR are able to produce large amounts of their electricity from hydroelectric power.

Tide and wave energy

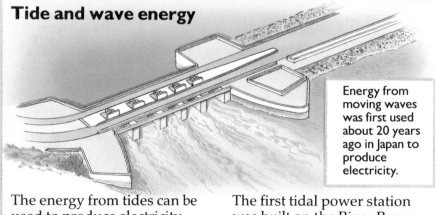

Energy from moving waves was first used about 20 years ago in Japan to produce electricity.

The energy from tides can be used to produce electricity. Tide water coming in is trapped behind a dam and then allowed to flow back through turbines.

The first tidal power station was built on the River Rance, France, in 1966. It provides enough electricity for a town of about 300,000 people.

Wind energy

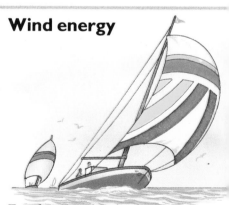

For thousands of years, wind energy has been used to push sailing boats and turn windmills. Today, windmills are used to produce electricity.

*Electricity, 102; Generators, 99; Kinetic energy, 11; Potential energy, 11; Water-wheels, 44.

Solar energy

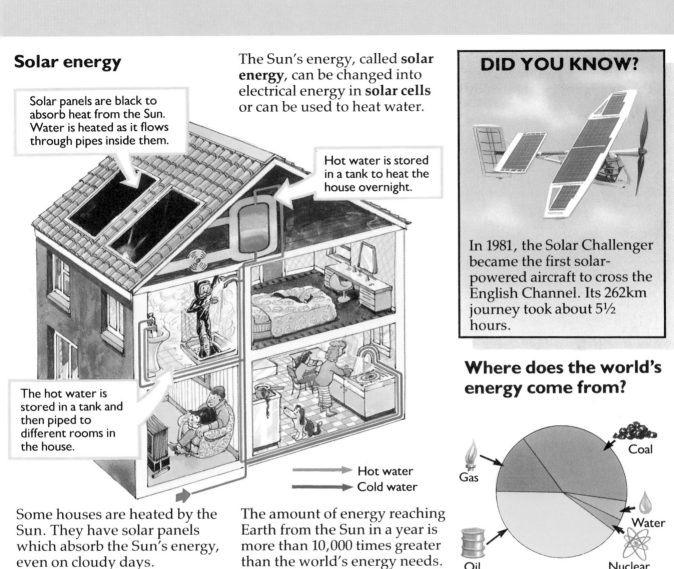

Solar panels are black to absorb heat from the Sun. Water is heated as it flows through pipes inside them.

Hot water is stored in a tank to heat the house overnight.

The hot water is stored in a tank and then piped to different rooms in the house.

The Sun's energy, called **solar energy**, can be changed into electrical energy in **solar cells** or can be used to heat water.

→ Hot water
→ Cold water

Some houses are heated by the Sun. They have solar panels which absorb the Sun's energy, even on cloudy days.

The amount of energy reaching Earth from the Sun in a year is more than 10,000 times greater than the world's energy needs.

Where does the world's energy come from?

Gas

Coal

Water

Oil

Nuclear

Windmills do not cause pollution, but they are large and noisy. To provide large amounts of energy, they take up huge areas of land.

Geothermal energy

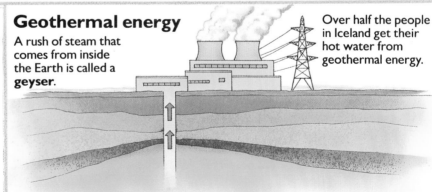

A rush of steam that comes from inside the Earth is called a **geyser**.

Over half the people in Iceland get their hot water from geothermal energy.

The inside of the Earth is very hot. It gets about 3°C hotter every 100m further down you go. In some places, especially near faults*, boiling water or steam rushes to the surface. This kind of energy, called **geothermal energy**, can be used for heating and to produce electricity.

Why do things move?

Nothing moves by itself. Things only move when they are pushed or pulled. Something which pushes or pulls is called a force. If there are no forces pushing or pulling, objects stay still or keep on moving at a steady speed in the same direction. There are many types of force.

Things that float are held up by a force called the **upthrust**.

Some metals are pulled towards magnets by a **magnetic force**.

A force called **gravity** pulls everything down towards the Earth.

Forces can make objects speed up, or **accelerate**. The larger the force, the more the objects speed up.

To stretch a bow, you have to pull against an **elastic force**, also called **tension**.

A drop of water is held together by a force called **surface tension**.

Measuring forces

Forces are measured in units called **newtons (N)**, with a spring balance. The force stretches the spring. The bigger the force, the more it stretches.

You push against an **elastic force** when you push a spring together.

When you move two surfaces against each other, a force called **friction** works to slow them down.

What can forces do?

A force can make things move, or it can change the speed of something that is already moving.

A force can change the direction of something that is moving.

Make an electrical force

Comb your hair with a plastic comb. You can then use it to pick up tiny bits of paper. Moving the comb through your hair gives it an electrical force that pulls paper to it.

A force can change the shape of something.

Joining and balancing forces

In a tug of war, the force of each person in a team pulling in the same direction adds up to make a bigger force.

When the force of both teams pulling in opposite directions is balanced equally, neither team moves.

When one team pulls harder than the other, the forces are unbalanced. Then both teams move in the direction of the team that pulled hardest.

Unbalanced forces

When a bicycle moves at a steady speed on a flat road, the force moving it forward is balanced equally by the force of friction* pulling it back.

Friction slows the bicycle down.

When the forces are unbalanced, the bicycle changes speed.

The force of the cyclist's legs moves the bicycle forward.

When the cyclist pushes harder, the bicycle goes faster. The force pushing it forward is greater than the force of friction holding it back.

When the cyclist stops pushing as hard, the bicycle slows down. The force of friction slowing it down, is greater than the force of the cyclist pushing it.

Action and reaction

The harder she pushes the water back, the more she moves forward.

Forces always come in pairs. The swimmer pushes the water back, making her move forward. The force pushing backwards is called the **action**. The force pushing forwards is called the **reaction**. Every action has an equal and opposite reaction. This means that whenever one thing puts a force on another, a force of equal size works in the opposite direction.

DID YOU KNOW?

Not all the cannons on one side of a 16th century sailing ship could be fired at once. The action would have caused such a big reaction that the ship would have capsized.

Friction

If you try to push a book gently along a table, at first it will not move. This is because a force called friction holds it back. As you push harder, the book eventually begins to slide.

But as the book rubs along the table, the force of friction slows it down. Friction always works to stop things moving, or to slow them down when they are moving.

No surface is ever completely smooth. Even something that looks smooth, like metal, looks bumpy through a microscope.

There is more friction on rough surfaces than smooth ones. When you write, friction makes the pencil lead rub off on to the

paper. But try writing on glass. Glass is smoother than paper, so there is less friction, and the pencil will not write well.

Friction can be a help

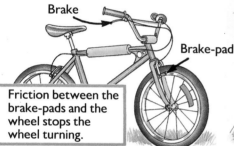

Brake

Brake-pad

Friction between the brake-pads and the wheel stops the wheel turning.

If a road is icy, there is less friction, so car tires have less grip.

Brakes work because of friction. The harder you squeeze them, the more the brake-pads push against the wheels and the quicker you stop.

Boots that climbers wear have rough, rubber soles. The friction between them and the rock stops the climbers' feet from sliding.

Roads and car tires are made with rough surfaces so that there is as much friction as possible between them. This is what stops cars skidding.

Friction is a drag

Trucks use more fuel than cars. Their heavy weight pushes their wheels harder on to the road, causing more friction.

Much of the fuel that cars use is wasted pushing against the force of friction.

DID YOU KNOW?

There is always friction between the moving parts in a machine. Machines need extra energy and use up more fuel as

they push against the force of friction. Because the parts of a machine rub against each other, they eventually wear out.

When you rub your hands together, the heat you feel comes from friction. The harder you rub, the warmer they get. The energy used to push against friction changes into heat. This is why machines are hot after use.

Getting rid of friction

A **lubricant**, like oil, reduces friction.

Smooth surfaces are easier to dance on than rough surfaces, because there is less friction.

Putting a thick liquid, like oil, on the moving parts in a machine stops them rubbing against each other. This cuts down friction which saves energy and stops the machine wearing out.

Rolling over friction

After wheels had been invented, they were used instead of logs.

Thousands of years ago, people found that it was easier to move heavy loads by rolling them on logs, than by dragging them along the ground. Rolling causes less friction than dragging.

Friction in the air

Things with a smooth shape are called **streamlined**.

The friction between anything moving and the air around it is called **air resistance**. How much air resistance there is on something depends on its shape. Cars are designed so that air moves smoothly over them, to cut down air resistance.

Ball-bearings

Another way of reducing friction inside machines is to use **ball-bearings**. These are small balls that roll over each other, keeping the moving parts of the machine apart.

Red-hot friction

There is no air in Space, so there is no friction to slow things down. Spacecraft only use their engines now and then so that they can change course.

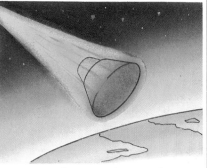

Spacecraft enter the Earth's atmosphere* so fast that they glow red-hot. This is because there is so much friction between them and the air.

A cushion of air

Hovercraft can travel over ground as well as over the water.

The world's largest hovercraft can carry over 400 passengers and 60 cars, and travel at a speed of 120km per hour.

Boats and submarines have to push through water. Friction between them and the water slows them down. **Hovercraft** work by floating on a cushion of air. This keeps them apart from the dragging force of the water. There is so much less friction that they can move much faster than ordinary boats.

Gravity

If you let go of something, it falls downwards. There is an invisible force called gravity pulling everything towards the Earth.

Without gravity, things would not be held on to the Earth's surface. They would fall off the Earth and go into Space.

What does gravity do?

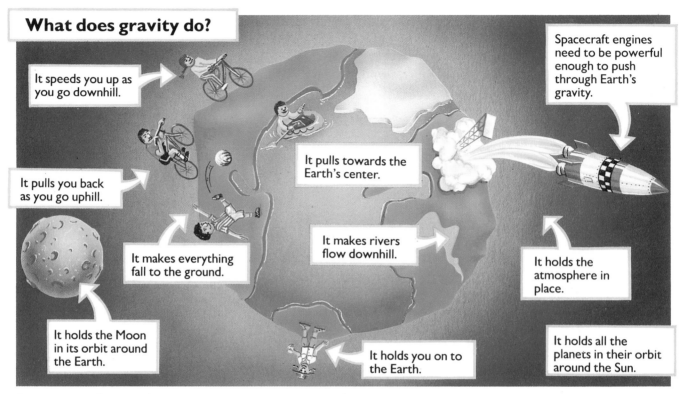

It speeds you up as you go downhill.

It pulls you back as you go uphill.

It makes everything fall to the ground.

It holds the Moon in its orbit around the Earth.

It pulls towards the Earth's center.

It makes rivers flow downhill.

It holds you on to the Earth.

Spacecraft engines need to be powerful enough to push through Earth's gravity.

It holds the atmosphere in place.

It holds all the planets in their orbit around the Sun.

Gravity was first understood about 300 years ago by Isaac Newton. Gravity is a force that attracts every object to every other object. You only notice the pull of gravity with things that are very large, like the Earth.

When you weigh something, you are measuring the force of gravity pulling it towards Earth. The further from the center of the Earth you go, the smaller the pull of gravity. So things weigh slightly less on top of high mountains.

Balancing things

Center of gravity

Center of gravity

Center of gravity

All things have a point, called their **center of gravity**, where their weight balances. A tray will only balance if you hold it under this point. Things that are top-heavy have a high centre of gravity. This makes them fall over more easily.

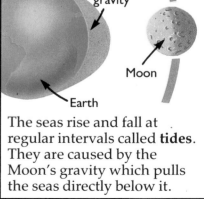

Changing gravity

The strength of gravity is different on other planets than it is on Earth. So the **weight** of an object changes if you take it to another planet. But the amount there is of the object, its **mass**, stays the same wherever it is.

Like all forces, weight is measured in **newtons (N)**. But when you weigh something, you really want to know how much there is of it, that is, its mass. Weighing scales measure the weight, and then convert the answer into units of mass, such as kilograms*.

On Earth, this astronaut has a weight of 600N and has a mass of 60kg.

On the Moon, his weight would be 100N, but his mass would still be 60kg. This is because the Moon's gravity is only 1/6 as strong as Earth's.

Jupiter's gravity is 264 times stronger than Earth's. There, his weight would be 158,400N. His mass would still be 60kg.

Quiz
On Earth 1kg weighs about 10N. What is your weight in N? What would your weight be on the Moon? (See page 128.)

Falling and gravity

About 400 years ago, Galileo noticed that objects speed up, or **accelerate**, as they fall.

He found that heavy and light things, of the same shape and size, take the same time to fall to the ground. Gravity pulls them down equally.

Try this for yourself with different objects, like a slipper and a heavy boot.

Air resistance

Objects of different shapes and sizes fall at different rates. The shape of a parachute makes people fall more slowly.

The air pushing against the parachute causes a lot of air resistance*. A person without a parachute falls faster because less air pushes against them.

The faster something falls, the more air resistance there is on it. Eventually, the air resistance slowing it down becomes as strong as the pull of gravity. Then its speed no longer changes. This is called **terminal speed**.

Free fall

If there were no air, there would be no air resistance. All falling objects would just get faster and faster at the same rate. This is called **free fall**.

Going straight

Things move because a force pushes or pulls them. Once moving, things only slow down, speed up or change direction if a force makes them. If no forces pushed on a moving object, it would carry on moving forever at the same speed in the same direction.

Speeding up is called **accelerating** and slowing down is called **decelerating**.

The heavier things are, the bigger the force needed to make them accelerate.

The force of friction* in the brakes makes cars decelerate. Fast cars need strong brakes to decelerate quickly.

The force of the engine pushes the car forwards. The more powerful the engine, the higher the acceleration.

How far something moves in a certain time is called its **speed**. You measure speed by counting how many metres something moves every second (**m/s**), or how many kilometres it moves every hour (**km/h** or **kph**). How fast something is going, but in a *particular direction*, is called its **velocity**. The velocity of a racing car may be 150kph towards North, and its speed 150kph.

Getting things moving ... and stopping them

Things with a large mass* have more inertia than things with a small mass.

Things that are not moving prefer to stay still, and things that are moving prefer to carry on moving. This is called **inertia**. Everything has inertia, and the bigger its mass, the more inertia it has. When a bus starts moving, you feel yourself being pulled backwards because your body's inertia makes it want to stay still. When the bus stops, you feel yourself being pulled forwards because your body's inertia makes it want to carry on moving at the same velocity.

Friction, 30; Mass, 33.

Inertia tricks

Put a plastic glass of water on a sheet of paper on a table. Then pull the paper away, very quickly and firmly.

Use an unbreakable glass.

The glass and paper must be dry.

The glass stays where it is because of its inertia. This only works if you pull the paper quickly enough.

Liquids have inertia too

You can tell the difference between a raw egg and a hard-boiled egg using inertia. Spin both eggs on saucers. As they spin, catch them and let them go again as quickly as you can. The boiled egg stays still. But the raw egg will start to spin again because of the inertia of the liquid inside it.

Bumps and blows

The force of hitting a ball makes it move. The ball then carries on by itself. Once any object is moving, it carries on moving with what is called its own **momentum**.

The harder you hit a ball, the more momentum it has and the further it goes. The lighter the ball, the less momentum it has. A ping-pong ball has less momentum than a baseball.

If a moving ball bumps into another ball, the momentum of the first makes the second one move. When you catch a ball, the momentum of the ball makes you move too, but only slightly because you are so much heavier than the ball.

When you jump up and down, your momentum makes the Earth move. But because the Earth is 100,000,000,000,000,000,000,000 times heavier than you are, the movement is very, very tiny and you do not notice it.

DID YOU KNOW?

The fastest land animal in the world is the cheetah, which can run faster than 100kph.

It can accelerate from 0 to 70kph in about two seconds, which is faster than most cars.

Round the bend

Moving in a circle is different from moving in a straight line*. All things move in a straight line unless another force makes them change direction. When something is moving around a corner, it is changing direction the whole time, so there must be a turning force keeping it moving in a circle. That turning force is called a centripetal force.

When you swing something around, the centripetal force keeping it in a circle comes from your arm.

When you let go, the pulling force stops. Whatever you are turning will then move off in a straight line.

Without a centripetal force to keep her moving in a circle, the skater would move off in a straight line.

Spinning water

If you spin a bucket full of water around fast enough, the water will not fall out. The centripetal force keeps the water moving in a circle.

The centripetal force comes from the bottom of the bucket pushing on the water. If you do not swing it fast enough, the water falls out.

Going round a bend

The force keeping you driving around in a circle comes from friction* between the tires and the road. Racetracks have bends built on sloping banks.

The sloping bank helps to keep the bicycles going round the corner. They can go round the corner faster because the slope stops them skidding off in a straight line.

Spinning things dry

Washing machines spin wet clothes to get rid of water. The centripetal force comes from the drum pushing on the clothes, keeping them moving in a circle. Water is able to get through the holes, so it flies off in a straight line.

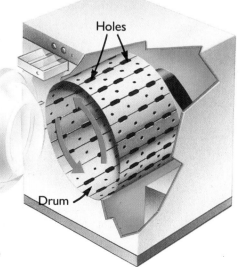

Holes

Drum

Keeping moving around a corner

Moving too fast to turn corner.

The faster something is moving, the more force is needed to keep it moving in a circle or around a bend.

Too heavy to turn corner.

In the same way, the larger the mass*, the bigger the force needed to keep it moving in a circle.

This corner is too tight to go around quickly.

A larger force is needed to keep things moving in a small circle than in a large one. So tight bends can be dangerous.

Tight corners

The friction between you and a car seat keeps you moving around a gentle curve. But if you go around a tight corner, you slide across the seat. This is because the force of friction is not strong enough to hold you.

DID YOU KNOW?

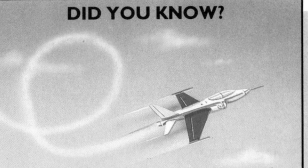

When pilots loop-the-loop, the centripetal force turning them can be so strong that they feel about four times heavier than normal.

Momentum

Things that are moving in a circle have momentum* just like things which move in a straight line. Momentum keeps a spinning top standing up. When it stops spinning it has no momentum so it falls over.

*Mass, 7, 33; Momentum, 35.   37

Floating and sinking

Mark the water level on a container full of water. Then add stones to the water and see how the water level rises. The water is pushed out of the way, or displaced, by the stones.

In water, things feel lighter than they really are. The water pushes on them, holding them up. When they are out of the water, they feel heavy again because the water no longer holds them up.

The bigger things are, the more water they displace, and the harder the water pushes back on them. The pushing force of a liquid is called **upthrust** or **buoyancy**.

Why does an iron ship float?

Solid iron is very dense, so even a small piece is very heavy. It sinks because the upthrust of water is not strong enough to hold it up. But ships are not just made of solid iron. They are hollow, full of big spaces filled with air.

Ships contain many spaces which are filled with air.

The air inside the ship makes it less dense than water.

The more water that something pushes out of the way, the harder the water pushes back on it.

The ship is held up by the upthrust of the water.

Why do some things float?

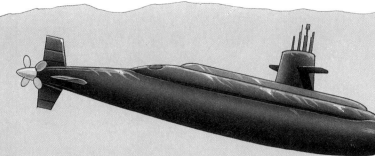

A piece of cork floats in water, but a piece of iron of the same size sinks. They displace the same amount of water, because they are both the same size.

The cork floats because it is much lighter for its size than the iron. How heavy something is for its size is called its **density**.

If something is less dense than water, it will float. This is because the upthrust of the water is strong enough to hold it up.

Submarines are able to change their density. When they fill their tanks with air, they float. When they fill their tanks with water, they sink.

Even though a ship may be very large, the air inside it makes it light for its size. The ship pushes so much water out of the way, that the upthrust of water pushing on it is strong enough to hold it up, making it float.

When a ship is being loaded with cargo, the plimsoll lines are checked to make sure that the ship does not float too low in the water.

If a ship is too heavily loaded, it becomes more dense than water and sinks.

These are called **plimsoll lines**, named after their inventor, Samuel Plimsoll.

Plimsoll lines show where the water level should come to in different sea conditions.

Walking on water

The surface of water has a sort of skin which is strong enough for tiny insects, like pond skaters, to walk on. This skin is called **surface tension**. It is surface tension that holds water together in drops.

Soapy water

When you add soap to water, you reduce its surface tension. This makes the water's skin stretchier. That is why you can make bubbles with soapy water.

Measuring upthrust

About 2,200 years ago, Archimedes noticed that water was pushed out of the way while he was getting into his bath. He found that the weight of water pushed out of the way is equal to the force pushing up on a floating object.

Anything can float

Things can float in any liquid or gas, as they do in water. Balloons float in air because they are less dense than air. Put some drops of cooking oil in water. The oil floats because it is less dense than water.

Salty water

Salt water has a greater density than fresh water, so ships float higher in salty water than in fresh water.

You can see this in the next experiment. Dissolve about 10 teaspoons of salt in a glass of warm water. Pour fresh water in another glass. Put an egg in each glass. The egg in the fresh water sinks, but the egg in the salty water floats.

DID YOU KNOW?

In the Dead Sea, the water is so salty that people can float in it without swimming. They can even sit in the water and read a book.

Pressure

The larger the area, the smaller the pressure.

The smaller the area, the larger the pressure.

Your feet sink into snow, unless you spread your weight over a larger area by wearing skis or snow-shoes. The push of your weight is then less on each point of the snow. The force pushing on a certain area is called pressure.

The girl's heels put more pressure on the ground than the elephant's feet, even though she weighs less. Why do you think sharp knives cut better than blunt ones? Why do nails have sharp points? (Answers on page 128.)

Pushing liquids

Fill a balloon with a little water and tie the end. Squeeze it between two plastic glasses. You cannot squash the water into a smaller space.

Liquids* cannot be squashed, so when you push on one part of a liquid, pressure is carried to all other parts of it.

In a car, pushing the brake pedal pushes liquid down a tube to the brakes. Because the liquid cannot be squashed, it carries the pressure from the pedal to the brakes.

Car disc brakes

Brake pedal

Pushing the brake pedal pushes liquid down tubes to the brakes.

Liquid

The liquid carries the pressure equally to the brakes on each wheel.

The liquid pushes the brakes on to a disc fixed to the wheel.

Pushing gases

Blow a little air into a balloon and tie up the end. Squeeze it between two glasses. Unlike water, you can squash the air into a slightly smaller space.

Gases* can be squashed, or **compressed**, into a smaller space. A compressed gas, like air in a balloon, pushes out equally in all directions. The more you compress a gas, the higher the pressure inside it.

Divers breathe in compressed air from metal bottles.

The bottles are very strong so they can hold the compressed air inside.

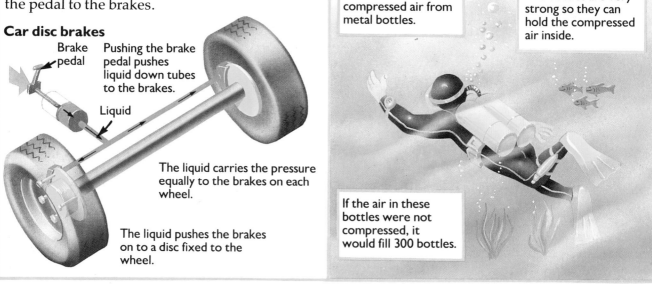

If the air in these bottles were not compressed, it would fill 300 bottles.

Pressure in liquids

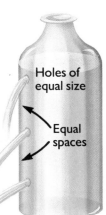

Holes of equal size

Equal spaces

Make three holes in a plastic bottle and cover them in sticky tape. Fill the bottle with water and then take off the tape.

Water from the bottom hole will squirt out furthest, because the weight of the water on top pushes on the water below it. The deeper the water, the greater the pressure.

Submarines are very strong so they are not crushed by the huge pressure deep in the sea.

Pressure in the air

Pressure in the atmosphere* works just like it does in water. The weight of the air above pushes on the air beneath it. This is called **atmospheric pressure**. The closer you get to the ground, the greater the atmospheric pressure.

Air pressure is measured with a **barometer**.

Changes in air pressure affect the weather*.

Pressure balance

Things do not collapse from the atmospheric pressure pushing down on them. This is because they are full of air which pushes out as hard as the air outside pushes in.

Your body is made so that you do not feel atmospheric pressure.

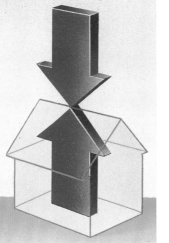

Liquid levels

Look inside a coffee pot or a teapot. The liquid inside the pot and the spout are always at the same level. This is because atmospheric pressure is pushing down evenly on both sides.

How does suction work? Pumps

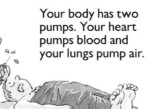

Your body has two pumps. Your heart pumps blood and your lungs pump air.

When you push a suction pad on to something, you push some air out, so the air pressure inside the pad is less than outside. The pad is held in place by the push of atmospheric pressure.

Pumps are used for moving liquids and gases around. A syringe is a simple pump. Pushing the plunger in increases the pressure inside, so the liquid is pushed out.

DID YOU KNOW?

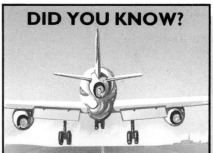

Air pressure changes as you change height. In a plane, this can make your ears feel blocked. Yawning or swallowing lets air in or out of your ears, making the pressure in them the same as outside.

Simple machines

Thousands of years ago, people did everything using only their own or their animals' muscle power. Over the years, they invented machines to help make work easier. Work can mean many things, but in science, doing work means using a force to move an object.

Long ago, people found that it was easier to drag heavy loads by rolling them along logs. Later, they invented **wheels**.

They found that it was easier to split logs and stones by hammering in a triangular shaped piece of wood. This is called a **wedge**.

Lifting loads

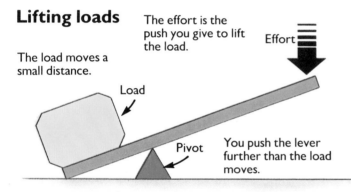

The effort is the push you give to lift the load.

The load moves a small distance.

Load

Effort

Pivot

You push the lever further than the load moves.

It is much easier to move a heavy load using a long stick called a **lever**. The lever is propped up on an object called a **pivot**. You have to push the lever further than you would have to push the load, but this is less effort than moving the load directly. A wheelbarrow is a type of lever. Its wheel works like a pivot. Scissors and shears are levers. The pivot is where the blades cross.

DID YOU KNOW?

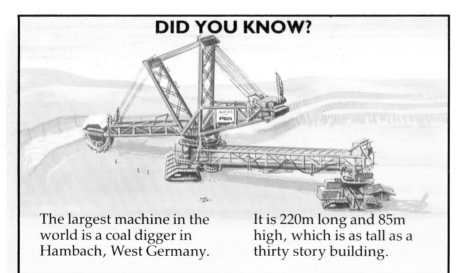

The largest machine in the world is a coal digger in Hambach, West Germany.

It is 220m long and 85m high, which is as tall as a thirty story building.

Try using a lever

Push down here to lift lid.

Long levers make work easier. Pry the lid off a can with a coin. Then try with a spoon. It is less effort with the spoon because it is a longer lever.

Slopes and screws

Vertical ladder

Sloping plank

A **slope** makes lifting loads easier. Although you have to go further, it is less effort to carry a load up a gentle slope than to lift it straight up.

A spiral staircase works like a coiled up slope. It is easier to walk up the stairs than to climb straight up, but you have to walk further.

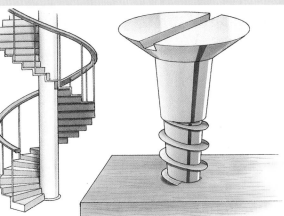

A **screw** works like a spiral staircase. You have to turn a screw round and round to get it in a wall, but this is easier than pushing it straight in.

Pulleys

A **pulley** helps you lift things. Pulling down with a pulley is easier than pulling things up, because your weight helps you.

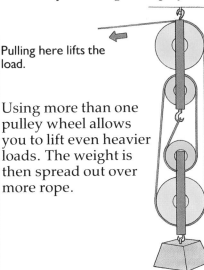

Pulling here lifts the load.

Using more than one pulley wheel allows you to lift even heavier loads. The weight is then spread out over more rope.

Archimedes' screw

The turning screw carries up water.

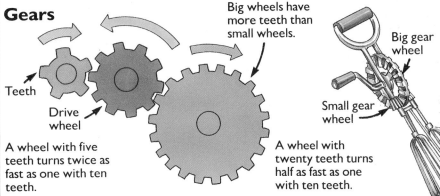

The handle turns the screw.

This machine was built to draw up water using a screw. It was invented by Archimedes in Greece, about 2,200 years ago.

Gears

Big wheels have more teeth than small wheels.

Teeth

Drive wheel

Big gear wheel

Small gear wheel

A wheel with five teeth turns twice as fast as one with ten teeth.

A wheel with twenty teeth turns half as fast as one with ten teeth.

Gears are toothed wheels. They are used to change speed. When one wheel, the drive wheel, goes round, it turns the one next to it. The drive wheel can make a smaller wheel turn faster, or a larger wheel turn slower.

You can see this with an egg beater. The big gear wheel turns when you turn the handle. This makes the small wheel and the egg beater connected to it spin much faster than you could turn them by hand.

43

Engines

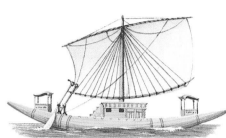

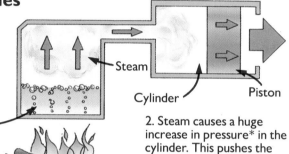

At first, people worked their simple machines by hand or with the help of animals. Then they learned to use the power of wind to push their sailing-boats along. They also used the wind to turn sails to grind grain into flour.

Later, they used the power of moving water in rivers to turn water-wheels. These pumped water or worked machinery.

Steam engines

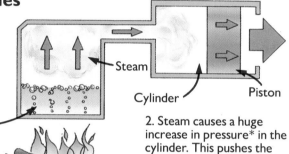

1. Coal or wood is burned to heat water. The water is boiled and changes into a gas, called **steam**.

Water

Fuel

Steam

Cylinder

Piston

2. Steam causes a huge increase in pressure* in the cylinder. This pushes the piston out.

3. Steam takes up about 1,700 times more space than the water it comes from.

The first engine invented to work a machine was a **steam engine**. Steam engines change the heat from burning fuel into movement.

The steam age

The steam engine was invented in 1777. Steam power was soon used to work many machines, and people moved into towns to work in new factories. The time when this was happening is called the **Industrial Revolution**

Steam locomotives

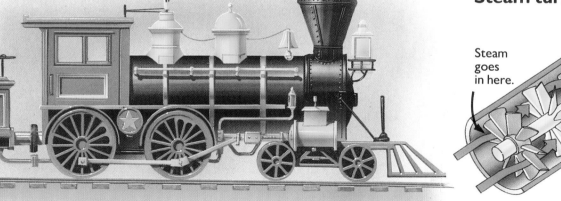

Later, steam engines were fitted on to wheeled carriages that ran on rails. These were called **steam locomotives**.

The first passenger railway opened in 1825 in Britain. Just 100 years later, railways had spread all over the world.

Steam turbines

Steam goes in here.

Spinning turbine blades

Today, steam power is used in power stations*. Steam pushes the blades of a turbine around, producing electricity.

Cars

Until the steam engine was invented, people rarely travelled far. They either rode on horse-back or in a horse-drawn carriage.

The first car, built in 1769, had a steam engine. Steam cars were slow and dirty. Their engines were large and heavy, and they carried a lot of fuel.

The first successful cars were built in Germany by Daimler and Benz in 1885-86. They used a new type of engine, called an **internal combustion engine**.

Internal combustion engines

Nikolaus Otto built the first **internal combustion engine** in 1876. It was smaller than a steam engine. It used a new fuel, **gasoline**, which wa light and easy to carry.

> ### DID YOU KNOW?
>
> There are about 300 million cars in the world today. That is one car for every 15 people. Every year, about 30 million cars are made.

How does a car work?

A **diesel engine** is an internal combustion engine that uses diesel fuel. Hot air makes the fuel explode, instead of an electric spark.

1. To start a car, the driver switches on an electric motor for a short time. This starts the pistons moving.

3. As this piston moves up, it squashes the mixture into a small space.

2. As this piston moves down, it sucks in a mixture of gasoline and air.

6. Water, cooled by air blowing on the radiator, is pumped around the engine to keep it cool.

4. An electric spark makes the fuel and air mixture explode, pushing this piston down.

5. This piston moves up and pushes waste gases out to the exhaust pipe.

7. Gears link the wheels to the engine. Different gears make the wheels turn at different speeds from the engine.

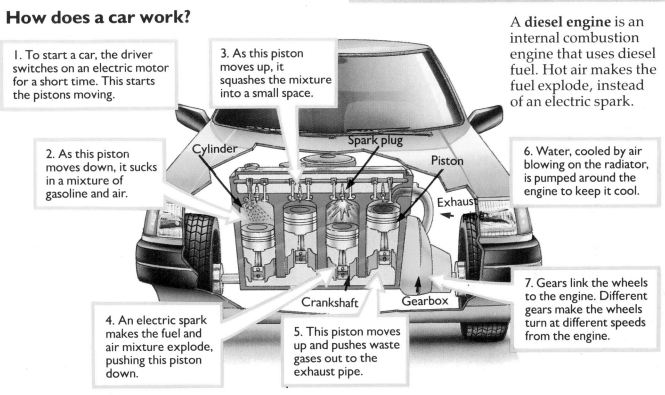

Cylinder · Spark plug · Piston · Exhaust · Crankshaft · Gearbox

The mixture of gasoline and air is exploded inside **cylinders**. This is what makes the pistons move up and down.

The **crankshaft** turns the up and down movement into a round and round movement, which turns the wheels.

Most engines are **four-stroke engines**. This means that, at any one time, each cylinder does one of the four jobs shown above.

Things that fly

Hot-air balloons are able to fly because they can float* up in the air. Airplanes are too heavy to float in air. They fly because they have wings. Their wings provide a force, called lift, that holds them up.

Aileron

The **ailerons** make the plane roll from side to side.

Raising the **spoilers** reduces lift.

The engine pushes the plane forwards, so that air flows over the wings.

Rudder

The **rudder** makes the plane turn to the left or right.

Spoiler

Lowering the **flaps** increases lift.

The **elevators** make the plane dive and climb.

Elevator

How do wings work?

To see how wings work, blow hard over a strip of paper, and watch the paper rise.

The faster air flows, the lower its pressure*. So as you blow, the pressure under the paper becomes greater than above it. This pushes the paper up.

The force pushing the wing up is called **lift**.

Airflow

The shape of a wing is called an **aerofoil**. It is designed so air flows faster over the top of it. This lifts the plane up.

Jet engines

Most new airplanes that are built today have **jet engines**. To see how they work, blow up a balloon and let it go. Air rushing out pushes the balloon forwards.

1. The **compressor blades** spin very fast, sucking air into the engine.

3. Gases shoot out, pushing the engine forwards and spinning the **turbines**.

2. Kerosene fuel explodes in the **combustion chamber**, making very hot gases.

4. The turbines are connected to the compressor. They keep it turning and sucking in air.

*Floating, 38; Pressure, 40.

Helicopters

Helicopters have rotor blades instead of wings. The blades are shaped like aerofoils. When the blades spin, the helicopter takes off.

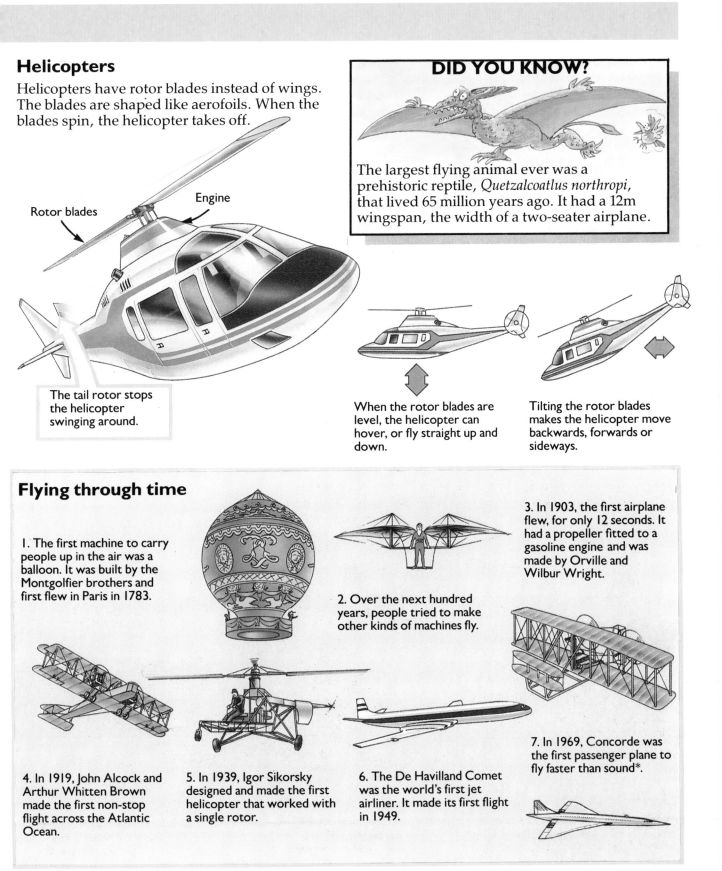

Rotor blades

Engine

The tail rotor stops the helicopter swinging around.

DID YOU KNOW?

The largest flying animal ever was a prehistoric reptile, *Quetzalcoatlus northropi*, that lived 65 million years ago. It had a 12m wingspan, the width of a two-seater airplane.

When the rotor blades are level, the helicopter can hover, or fly straight up and down.

Tilting the rotor blades makes the helicopter move backwards, forwards or sideways.

Flying through time

1. The first machine to carry people up in the air was a balloon. It was built by the Montgolfier brothers and first flew in Paris in 1783.

2. Over the next hundred years, people tried to make other kinds of machines fly.

3. In 1903, the first airplane flew, for only 12 seconds. It had a propeller fitted to a gasoline engine and was made by Orville and Wilbur Wright.

4. In 1919, John Alcock and Arthur Whitten Brown made the first non-stop flight across the Atlantic Ocean.

5. In 1939, Igor Sikorsky designed and made the first helicopter that worked with a single rotor.

6. The De Havilland Comet was the world's first jet airliner. It made its first flight in 1949.

7. In 1969, Concorde was the first passenger plane to fly faster than sound*.

Space

Spacecraft are launched into Space by very powerful rocket engines. A rocket engine is the only kind of engine strong enough to push through Earth's gravity.

Rockets and rocket engines

Rocket engines work in the same sort of way as jet engines*. They move forward by pushing out a powerful stream of gases made by burning fuel.

Nothing burns without oxygen. As there is no oxygen in Space, rockets carry their own supply. They use liquid oxygen or an **oxidant**, which is a chemical containing oxygen, to burn fuel.

Oxidant

Fuel

Hot gases rush out, pushing the rocket forwards.

← Booster rockets

Fuel burns here

This rocket is called Ariane. It was built by the European Space Agency and is used to put satellites into orbit.

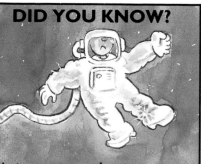

DID YOU KNOW?

Astronauts can become about 5cm taller while they are in Space. Their spines stretch out because gravity does not squash them down.

To escape the pull of gravity, booster rocket engines give the spacecraft an extra push at the beginning of a flight. They fall off when they have used up all their fuel.

Space milestones

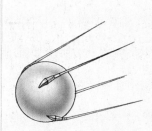

1957. Sputnik 1 (USSR) was the first spacecraft to orbit around the Earth. It was a satellite that weighed 84kg and was only 58cm across.

1961. Yuri Gagarin (USSR) was the first man in Space. He orbited the Earth for 108 minutes in a spacecraft called Vostock 1.

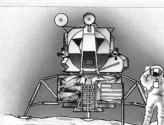

1969. Apollo 11 (USA) landed the first men on the Moon, the astronauts Edwin Aldrin and Neil Armstrong. Their first Moon walk lasted 2½ hours. They collected Moon rock and soil samples to study on Earth.

1976. Viking 1 (USA), an unmanned space probe, went to Mars. It tested soil samples and sent pictures back to Earth.

What is space like?

People have only travelled as far as our Moon. But unmanned spacecraft, called **probes**, have explored further into Space.

There is no air in Space. Something that is completely empty is called a **vacuum**. On Earth, things may look empty, but they are really full of air.

Spacecraft only need to use their engines to change speed or direction, because in a vacuum there is no air resistance* to slow them down.

A spacesuit protects the astronaut. Liquid pumped through it keeps the temperature steady. The pressure* inside the suit is kept the same as on Earth.

Astronauts carry their own supply of oxygen to breathe.

Sound needs something to travel through. Because Space is a vacuum, astronauts use radios to talk to each other.

Without an atmosphere to absorb temperature changes, it is hotter than an oven when you face the Sun, and colder than a freezer in the shade.

A spacecraft does not use its engines to keep in its orbit. It is kept moving because of Earth's gravity. The Earth's gravity pulls the spacecraft and the astronauts equally. But there is no force pulling the astronauts to the spacecraft, so they float in it and feel weightless.

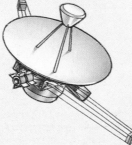

31. The space shuttle ...lumbia (USA) was ...nched. It was the first ...cecraft that could be ...d again.

1987. Pioneer 10 (USA) passed Pluto, the furthest planet from the Sun. It is the first man-made object to leave our Solar System.

A light year

Our nearest star is 4½ light years away. Its light takes 4½ years to reach Earth.

Distances on Earth are measured in metres. In Space, distances between stars are so large, that a bigger unit of length is used, called a **light year**. A light year is the distance light travels in one year, about 10 million million kilometres.

*Air resistance, 31; Pressure, 40. **49**

Light and dark

Light is a form of energy. Things that give out light of their own are called luminous. The Sun, electric light bulbs, candles and televisions are all luminous objects.

Things that do not give out light are lit up by luminous objects. The Earth's biggest source of light is the Sun. All living things on Earth depend on the energy from sunlight.

Light is the fastest thing in the Universe.

Light rays

Light travels in straight lines called **rays**. You can see this when you look at sunlight pouring in through a window or at the beam of a flashlight.

Light from the Sun takes eight minutes to travel the 150 million km to Earth.

Light travels 300,000km each second, which is a million times faster than a Jumbo jet.

Shadows

Things which light cannot pass through are called **opaque**. Most things are opaque. Shadows are formed on the other side of opaque objects, where light cannot reach.

Things which light can pass through, like glass, are called **transparent**.

Things which only a little light can pass through, like dark glasses, are called **translucent**.

A large source of light makes a shadow which is dark in the center, and lighter round the outside.

Light and dark shadows

There are different types of shadow. Dark shadow, called **umbra**, is formed where no light reaches. When some light gets through, the shadow looks grey and is called **penumbra**.

Shadows change depending on the size of the light source.

A small source of light casts a very dark shadow with sharp edges.

Umbra

Umbra

Penumbra

50

Shadow time

Shadows can give you some idea of the time of day. They are long in the morning and evening, and short in the middle of the day when the Sun is overhead.

Eclipses

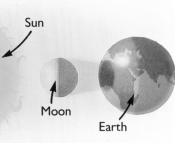

Sun

Earth

Moon

Sun

Moon

Earth

When the Moon, Earth and Sun are in a line, the Moon is completely covered by the Earth's shadow, so that you cannot see it. This is called a **lunar eclipse**.

A **solar eclipse** is when the Moon blocks out the Sun, so you cannot see it. The shadow of the Moon falls on the Earth. Solar eclipses happen less often than lunar eclipses.

The brightness of light

Light from some things is brighter than from others. The brightness of light is called its **intensity**.

Light spreads out, so the further you are from its source, the less intense it is.

Light from this lamp is more intense than candle-light.

DID YOU KNOW?

These are views of the Moon from the northern half of the Earth.

The Moon takes 27.3 days to orbit the Earth. During this time, it looks as if it is gradually changing shape.

This is because we see different amounts of the Moon's sunlit side as it moves around the Earth.

Bouncing light

A light bulb lights up a whole room because light from it bounces off everything in the room. In the same way, light from the Sun bounces off everything. Days are light because sunlight is bounced in all directions, or scattered, off tiny pieces of dust in the atmosphere. Things that give out light, like the Sun, are called luminous. Most things are not luminous. You only see them because light bounces, or is reflected, off them.

Light from the most distant star in our galaxy* takes about 80,000 years to reach Earth.

You can only see the planets of our solar system because they reflect the light of the Sun.

The Sun and all the other stars are the only luminous things in Space.

The Moon is not luminous. You see it because it reflects light from the Sun.

Space is dark because it is completely empty. There is no atmosphere to scatter light from the stars.

When a cloud blocks out the Sun, the sky does not become completely dark because the atmosphere scatters sunlight.

Days are light because sunlight is scattered in all directions by the Earth's atmosphere.

Scattering light

You can see how the atmosphere scatters light when you are at the movies. You see the beam from the projector because dust in the air reflects the light.

DID YOU KNOW?

In 1969, the exact distance from the Earth to the Moon was worked out from the time light took to travel there and back. Light from lasers* on Earth was reflected from a mirror that astronauts placed on the Moon.

*Galaxy, 22; Lasers, 109.

How light bounces

Light bounces just like a ball. When it hits a surface straight on, it bounces straight back. When it hits a surface at an angle, it bounces off at the same angle.

Rough and smooth

When light hits smooth surfaces, it is all reflected in one direction. When light hits rough surfaces, it bounces off in lots of different directions.

White reflects light, so it shows up better than black.

Like heat radiation*, light is reflected off some things better than others. White surfaces reflect more light than they absorb. Black surfaces absorb more light than they reflect.

Mirrors

The light that bounces off you is reflected back by the mirror.

Mirrors reflect light best because they are very smooth and shiny. The reflection you see is not the same as you. When you wave your right hand, your reflection waves its left hand. A reflection is the wrong way round.

Mirror writing

Look at writing in a mirror. The mirror turns all the letters the wrong way round so you cannot read them. You can put the mirror writing in this picture back the right way round if you look at it through a mirror.

Seeing with mirrors

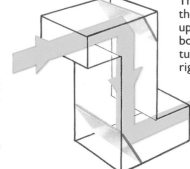

The reflection in the top mirror is upside down. The bottom mirror turns it back the right way up.

The crew of a submarine can look out from under the sea using a **periscope**. This is a long tube with two mirrors, one at each end.

Mirrors and pictures

Mirrors that bulge out

A curved mirror makes things look different. A mirror that bulges outwards is called a **convex mirror**.

Convex mirrors give a wide view. Side view mirrors are convex, to give drivers a wide view of what is behind them.

Mirrors that bulge in

A mirror that bulges inwards is called a **concave mirror**. How you look in one depends on how close you are to it.

From close up, your reflection looks larger, or is **magnified**. Further away, your reflection looks small and is upside down.

Curved mirrors

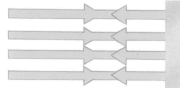

When light hits a flat mirror straight on, it all bounces straight back.

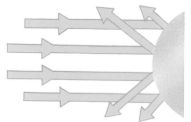

When light hits a convex mirror straight on, it bounces off at a wide angle.

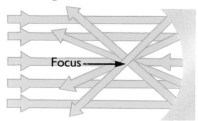

Focus ⟶

When light hits a concave mirror straight on, it bounces inwards and meets at a point called the **focus**.

Reflector telescopes

Concave mirrors are used in some telescopes for looking at the stars. The biggest in the world is on Mount Semirodriki, USSR. Its mirror is 6m across. With this telescope, you could see light from a candle 24,000km away.

Heating with mirrors

In Odeillo, France, a huge concave mirror is used to collect the Sun's rays. The rays bounce off and meet at the focus of the mirror, where the temperature is so high, about 4,000°C, that the heat can be used to melt metals.

Seeing far away

If you look down a street, the people in the distance seem much smaller than the people near you. But you know that people do not become smaller as they walk away.

This means you can tell how far away something is by its size. Because things look smaller further away, the road seems to get narrower until it comes to a point in the distance.

Painting what you see

Painted in Egypt, about 3,500 years ago

The earliest pictures people painted looked flat. Later, people began to paint things as they saw them.

Painted in Italy, about 650 years ago

They painted far away things smaller than things that were close by. This feeling of distance is called **perspective**.

Picture trick

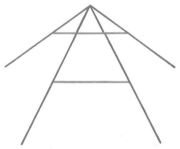

Your eyes can sometimes mislead you. Are both red lines the same length, or is the top one longer? (Answer on page 128.)

Bright lights

Light from a bulb spreads out, so that the light is less bright, or intense, further away. There are concave mirrors inside car headlights to stop the rays of light spreading out. This keeps the beam of light very bright even a long way off.

DID YOU KNOW?

Mirrors can be used to signal for help in an emergency. On a clear day, a beam of sunlight can be reflected off a mirror and seen up to 40km away.

Bending light

Whenever light passes through one transparent thing into another, it is refracted.

You see things around you because light bounces off them. If you look at something in water, like an oar, it looks bent. This is because light is bent, or refracted, as it passes from water to air.

The fisherman sees the fish here.

Because light bends, you see things in water in a different place from where they really are. To catch fish with a spear, people aim below the place where they see the fish.

Make light bend

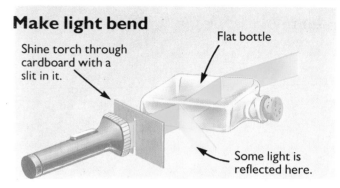

Shine torch through cardboard with a slit in it.

Flat bottle

Some light is reflected here.

Fill a bottle with water and a few drops of milk. In a dark room, shine a thin beam of light through it. The light is refracted by the water. When it passes from water to air on the other side, it bends back the other way.

Refraction also makes water look less deep than it is.

Light bends because it travels at different speeds through different things. It travels faster through air than through water, but faster through water than through glass.

Inner reflections

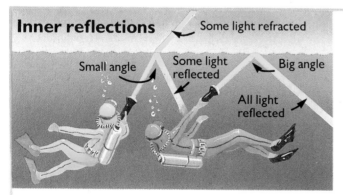

Some light refracted

Small angle

Some light reflected

Big angle

All light reflected

When light passes from water into air, some of it gets through and some is reflected. How much is reflected depends on the angle of the light rays. When all the light is reflected, this is called **total internal reflection**.

Pouring light

The water carries the light by total internal reflection.

Make a hole in a clear, plastic bottle. Holding your finger over the hole, fill the bottle with water. In a dark room, shine a flashlight from behind the hole and let the water pour out into a bowl. See how the water carries the light with it.

Lenses

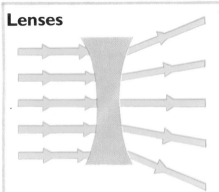

A **concave lens** is shaped to make light rays spread out. When you look through one, things look smaller.

A **convex lens** is shaped to refract rays of light so they meet. When they hit the lens straight on, all the rays meet at a point called the **focus**.

Making an image

The rays of light meet here.

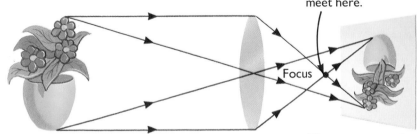

Focus

The image is upside down.

You can use a convex lens to make an image of something on a screen. The image looks sharp, or in focus, when the screen is at the place where all the rays of light meet. To find where the image is in focus, move the screen around.

Making things look bigger

A magnifying glass is a convex lens. When you hold it close to an object, it makes the object look bigger.

DID YOU KNOW?

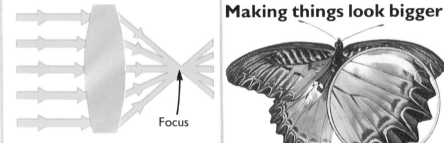

You should never leave bottles lying outside. They can act as lenses, focusing the Sun's rays on the ground, which could start a fire.

Optic fibers

The light is carried by optic fibers.

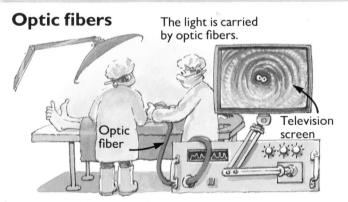

Optic fiber

Television screen

In the same way as water in the last experiment, thin rods of glass are used to carry light by total internal reflection. These are called **optic fibers**. Doctors use them to look inside people.

Mirages

On a hot day, you may think you see water far away. This is a **mirage**. You see a mirage when light from the sky is totally internally reflected off a layer of hot air near the ground.

Seeing pictures

How do your eyes work?

The things around you give out light or reflect light that hits them. You see them when that light enters your eyes.

The image on the retina is upside down. Your brain twists it round so things look the right way up.

A convex lens in your eye makes an image on the **retina**, at the back of your eye.

A transparent layer, called the **cornea**, protects your eye.

Your brain turns signals from the retina into the picture you see.

Light enters the eye through a hole called the **pupil**.

Pupil

Convex lens

Retina

Optic nerve

The colored part of your eye is called the **iris**.

These muscles change the shape of the lens.

When light hits the nerve cells in the retina, they send signals down the optic nerve to the brain.

Iris

The iris controls the size of the pupil.

These muscles move the eyeball.

Light and dark

The iris controls how much light gets into the eye. In the dark, the iris opens. This makes the pupil bigger to let in more light. In bright light, it closes up. Look in a mirror in a dark room and turn on a light. You can see your pupils change size.

Near and far

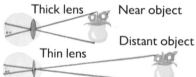

Thick lens   Near object

Distant object

Thin lens

The lenses in your eyes change shape when they focus on things that are near or far away.

Seeing clearly

Long-sighted people cannot focus on near things. Their glasses have convex lenses. Short-sighted people cannot focus on distant things. Their glasses have concave lenses.

Seeing with two eyes

Your eyes are a small distance apart so each sees a different view. This helps you see the shape of things and how far off they are. Most animals that hunt also have eyes facing forwards.

They can judge distances well, but have a narrow view. Other animals have eyes facing sideways. They have a wide view to look out for hunters, but cannot judge distances well.

DID YOU KNOW?

Eagles have the best eyesight of all animals, so that they can see their prey from very high in the sky. An eagle can spot a hare from 3km away.

The camera

A camera works like an eye. Light comes in through a lens. The lens makes an image on film at the back of the camera.

A camera must only let in light when you take a picture.

Different views

You can change lenses on some cameras. A **wide-angle lens** gives a wide view. A **telephoto lens** gives a close-up view so you can take pictures of something far away.

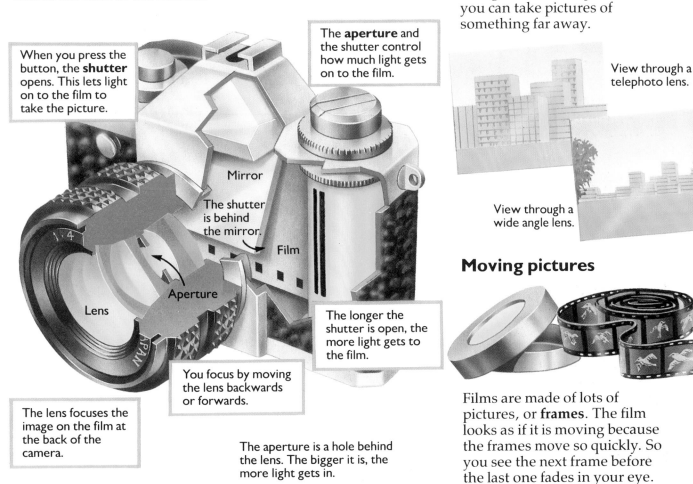

The **aperture** and the shutter control how much light gets on to the film.

When you press the button, the **shutter** opens. This lets light on to the film to take the picture.

Mirror

The shutter is behind the mirror.

Film

Aperture

Lens

You focus by moving the lens backwards or forwards.

The longer the shutter is open, the more light gets to the film.

The lens focuses the image on the film at the back of the camera.

The aperture is a hole behind the lens. The bigger it is, the more light gets in.

View through a telephoto lens.

View through a wide angle lens.

Moving pictures

Films are made of lots of pictures, or **frames**. The film looks as if it is moving because the frames move so quickly. So you see the next frame before the last one fades in your eye.

How a photograph is made

To take a photograph, you focus the lens of the camera and set the amount of light that goes in it. Automatic cameras are able do this for you.

Chemicals on the film change when light hits them. When you get the film developed, the picture appears on it. This is called a **negative**. On negatives,

dark things look light and light things look dark. Light is shone through the negative to make a picture on special paper. This is the print you see.

Colored light

About 300 years ago, Isaac Newton shone a beam of sunlight through a prism. He discovered that white light is made up of colors. Using a second prism, Newton found he could mix all the colors together to make white light again.

When light goes through a prism, it is bent, or refracted*, because the prism slows it down. White light is a mixture of colors. The colors travel at different speeds through the prism, so they are bent by different amounts.

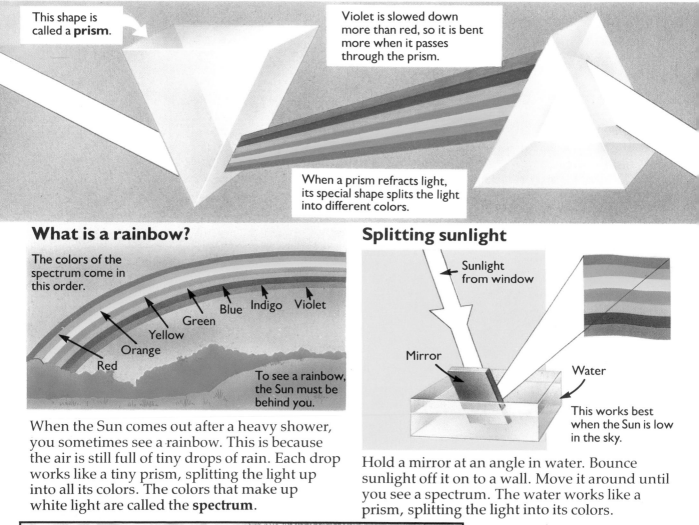

This shape is called a **prism**.

Violet is slowed down more than red, so it is bent more when it passes through the prism.

When a prism refracts light, its special shape splits the light into different colors.

What is a rainbow?

The colors of the spectrum come in this order.

Red
Orange
Yellow
Green
Blue
Indigo
Violet

To see a rainbow, the Sun must be behind you.

When the Sun comes out after a heavy shower, you sometimes see a rainbow. This is because the air is still full of tiny drops of rain. Each drop works like a tiny prism, splitting the light up into all its colors. The colors that make up white light are called the **spectrum**.

Splitting sunlight

Sunlight from window

Mirror

Water

This works best when the Sun is low in the sky.

Hold a mirror at an angle in water. Bounce sunlight off it on to a wall. Move it around until you see a spectrum. The water works like a prism, splitting the light into its colors.

Colors from diamonds

You can see the colors of the spectrum in a diamond. It is cut so that it reflects and refracts light, in the same way as a prism does.

DID YOU KNOW?

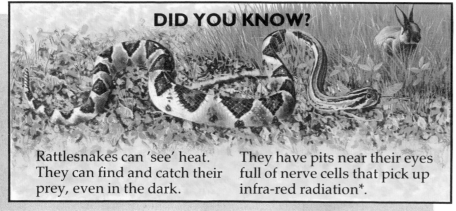

Rattlesnakes can 'see' heat. They can find and catch their prey, even in the dark.

They have pits near their eyes full of nerve cells that pick up infra-red radiation*.

*Infra-red radiation, 18; Refraction, 56.

What makes colors different from each other?

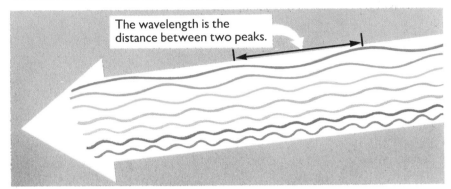

The wavelength is the distance between two peaks.

Heat radiation and light come to you from the Sun.

Light rays are made up of tiny waves, much too small to see. The size of the waves is measured by their **wavelength**.

Each color has a different wavelength. The wavelength of red light is longer than the wavelength of violet light.

Heat radiation* and light are similar. They both travel in waves, but they have different wavelengths.

Why is the sky colored?

Sunlight is scattered by the Earth's atmosphere. Some of the colors in sunlight are scattered more than others. The atmosphere scatters blue light most, so during the day, the sky looks blue.

At sunset, sunlight goes through more atmosphere to get to you. The blue light is then scattered so much that you never see it. The sky looks red because you see the scattered red light.

Seeing colors

There are special nerve cells in your eyes for seeing color. They only work well in bright light. This is why things look colorless in poor light.

Some color-blind people cannot see this number.

Colors give you a lot of information. Some people are color-blind. They cannot tell the difference between certain colors.

How animals see color

Not all animals see colors in the same way as people. A desert ant sees some colors better than you, but a squid cannot see colors at all.

*Heat radiation, 18.   61

Mixing colors

A blue filter only lets blue light through.

A red filter only lets red light through.

White light is made up of all the colors of the spectrum. You can divide it up into different colors with a color filter.

A filter is a piece of colored glass or plastic that makes colored light. It only lets one color through, blocking out all the rest.

Primary colors

Red, green and blue are called **primary colors**. They are special, because you can use them to make light of any color. If you mix any two primary colors, the colored light they make is called a **secondary color**.

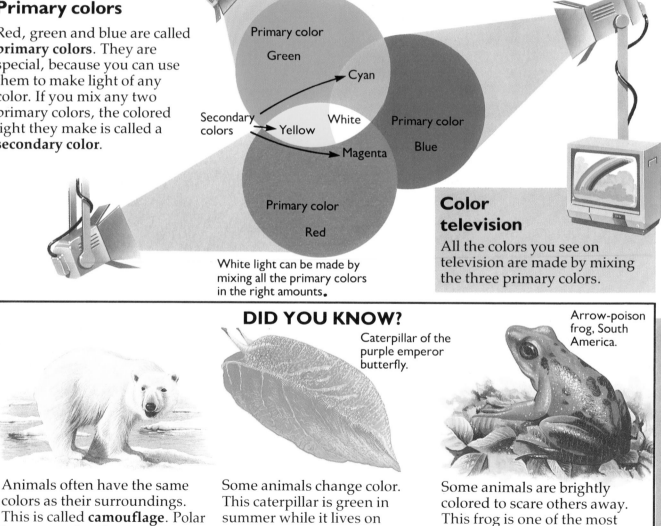

Primary color
Green

Cyan

Secondary colors

Yellow

White

Primary color
Blue

Magenta

Primary color
Red

White light can be made by mixing all the primary colors in the right amounts.

Color television

All the colors you see on television are made by mixing the three primary colors.

DID YOU KNOW?

Caterpillar of the purple emperor butterfly.

Arrow-poison frog, South America.

Animals often have the same colors as their surroundings. This is called **camouflage**. Polar bears are white because they live in the snow.

Some animals change color. This caterpillar is green in summer while it lives on leaves. But it turns brown in winter as it lives on twigs.

Some animals are brightly colored to scare others away. This frog is one of the most poisonous animals in the world.

Why do things look different colors?

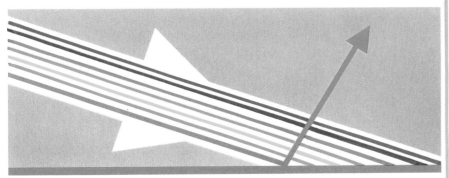

The color of something depends on the colors of light that it reflects. An object looks red because it reflects red light and absorbs all the other colors. Something looks blue because it reflects blue light and absorbs all the other colors. A white object reflects all the colors of light equally. But a black object reflects no light. It absorbs all colors.

Mixing paints

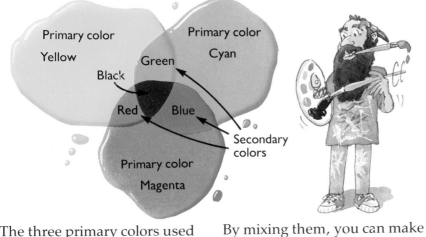

Primary color Yellow

Primary color Cyan

Green

Black

Red   Blue

Secondary colors

Primary color Magenta

The three primary colors used in painting are magenta, yellow and cyan. They are not the same as the primary colors for light.

By mixing them, you can make almost any color, apart from white. Mixing all three colors together makes black.

Changing colors

Things can look a different color when you see them in

colored light. A red dress will look black in blue or green light.

Printing colors

All the colors you see in this book have been printed using only four different colored inks. They are yellow, cyan, magenta and black.

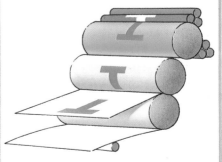

To print each page, the paper goes through the machine four times, each time with one of the different inks. This is called **four color printing**.

Black ink is used to make the picture darker.

Magenta ink   Cyan ink   Yellow ink

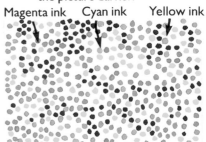

Look at the pictures on this page through a strong magnifying glass. They are made up of thousands of tiny dots of different colors.

Sound

The sounds you hear tell you what is happening around you, even if you cannot see what is making them. You may be able to hear the sound of a telephone ringing, of cars driving by, or of rain falling.

A sound is made when something moves backwards and forwards very quickly. This is called **vibration**. When something vibrates, it makes the air around it vibrate too. The sound you hear is carried by the vibrating air.

Sound is a form of energy, but the energy in most sounds is small. The sound energy of 200 pianos is equal to the electrical energy needed to light just one light bulb.

When you speak, air from your lungs makes vocal cords in your throat vibrate.

The sound of a violin is made by the vibrating strings.

The sound from radios and televisions comes from loudspeakers.

Electrical signals make the loudspeaker vibrate.

High and low sounds

The faster something vibrates, the higher the sound it makes. The slower it vibrates, the lower the sound it makes. How high or low a sound is, is called its **pitch**. The number of vibrations per second is called the **frequency** of the sound.

Frequency is measured in **hertz (Hz)**. Bees beat their wings 200 times a second, so the sound you hear has a frequency of 200 hertz. Mosquitoes make a higher pitched sound than bees because they beat their wings faster, about 500 times each second.

You can feel your vocal cords vibrating, if you touch your throat as you speak.

People and animals use sound to communicate with each other.

The speed of sound

Thunder and lightning happen at the same time, but you see lightning before you hear thunder, because sound travels much slower than light.

You can tell how far away a thunderstorm is. Count the number of seconds between seeing the lightning and hearing the thunder. Then divide the answer by three. This tells you how far away the storm is, in kilometres.

In one second, light travels 300,000km, but sound only travels 340m.

Concorde, the fastest passenger aircraft, can fly at twice the speed of sound.

You hear the sound of a crash because the collision makes the cars vibrate.

DID YOU KNOW?

Some airplanes are **supersonic**. This means that they can travel faster than the speed of sound. Their speed is measured in units called **mach**.

Mach 1 is equal to the speed of sound. The fastest jet in the world is the Lockheed SR-71 (USA) which flies at mach 3.5.

Sound can travel through liquids.

Sound can travel through solids.

Sound in solids and liquids

As well as travelling through air, sound can travel through liquids and solids. That is why you can hear sounds through walls and through water.

Far away sounds

The further away you are from the source of a sound, the quieter it gets. Because of this, you can sometimes tell how far away something is.

Sound travels

Sound moves

Sound travels in sound waves. They spread through the air like ripples in a pond after you have thrown in a pebble.

As a bell vibrates, it pushes and pulls the air around it, making layers of different air pressure*. This is a **sound wave**.

Each layer of air bumps into the next layer, carrying the sound to your ears.

Your ears pick up sound waves. The tiny differences in air pressure make your eardrums* vibrate in time with the sound of the bell.

Light waves* can travel through Space, but sound waves cannot.

Sound waves need something to travel through. Space is completely silent because there is no air to carry the sound.

Sound through gases

Sound waves can travel through gases. Most of the sounds you hear have travelled through air to reach you. Through air, sound travels about 340m each second.

Sound travels slightly faster in hot air and slightly slower in cold air.

Sound through solids

Sound can travel through solids. You can hear something, even if it is very far away, by listening to the ground.

Sound travels best through hard solids. It travels 15 times faster through steel than through air.

Sound through liquids

Sound waves can travel through liquids. You can hear the sound of someone splashing when you are swimming underwater. Sound waves always travel better through liquids than gases.

Sound travels about four times faster through water than through air.

Sound spreads out

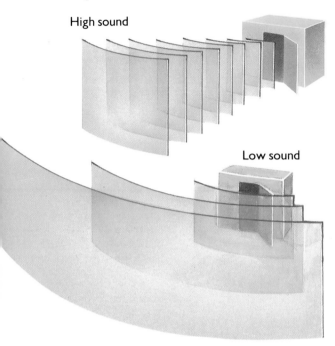

High sound

Low sound

You can hear around corners. This is because sound waves spread out as they go through gaps or around obstacles. This is called **diffraction**. Low sounds spread out, or are diffracted, more than high sounds. So, from far away, you hear the low notes in music better than the high ones.

Echoes

You can only hear sounds separately if they are over ¹/₁₀ second apart.

When you hear an echo, you are hearing sound waves reflected back to you from far away. You cannot hear echoes in small rooms because the walls are too close to you. Sound bounces back too quickly for you to hear the echo separately.

DID YOU KNOW?

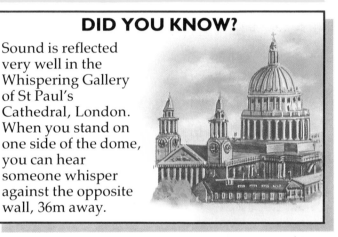

Sound is reflected very well in the Whispering Gallery of St Paul's Cathedral, London. When you stand on one side of the dome, you can hear someone whisper against the opposite wall, 36m away.

Sounds from a stage

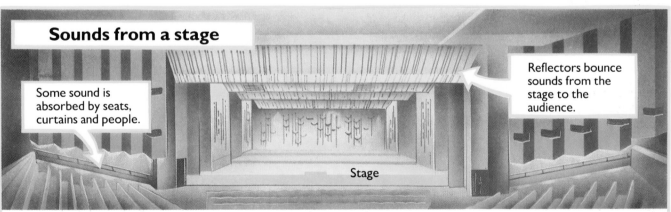

Some sound is absorbed by seats, curtains and people.

Reflectors bounce sounds from the stage to the audience.

Stage

The way sound travels in a room depends on its shape and what is in it. Hard, flat surfaces reflect sound waves well. Soft, bumpy surfaces absorb the sounds that hit them.

When sound waves meet, they can sometimes add up to make a louder sound, or cancel each other out to make a quieter sound. This is called **interference**.

Concert halls are built to avoid interference and echoes, so that sound carries well from the stage to the audience. The way sound travels in a room is called **acoustics**.

Hearing sound

Your ears pick up the vibrations made by sound waves. You hear sounds because nerves in your ears change the vibrations into signals that go to your brain.

As sound waves come into your ear, they make a sheet of skin, called the **ear-drum**, vibrate. The sound waves make the ear-drum vibrate in time with whatever made the sound.

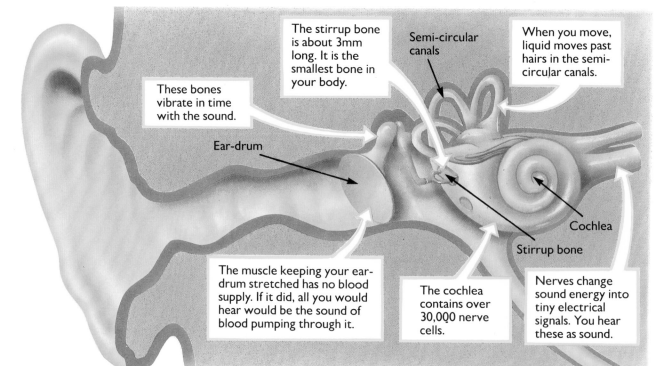

The stirrup bone is about 3mm long. It is the smallest bone in your body.

Semi-circular canals

When you move, liquid moves past hairs in the semi-circular canals.

These bones vibrate in time with the sound.

Ear-drum

The muscle keeping your ear-drum stretched has no blood supply. If it did, all you would hear would be the sound of blood pumping through it.

The cochlea contains over 30,000 nerve cells.

Nerves change sound energy into tiny electrical signals. You hear these as sound.

Cochlea

Stirrup bone

The ear-drum makes three tiny bones vibrate. They work like little levers, increasing the strength of the vibration. They are joined to a tube filled with liquid, called the **cochlea**.

The **stirrup bone** works like a piston, pushing liquid in the cochlea backwards and forwards, in time with the sound. Nerves change the vibration into electrical signals that go to the brain.

Your ears help you balance

The semi-circular canals in your ears help you balance. As you move, liquid inside them pushes on tiny hairs, sending nerve signals to the brain. You feel dizzy after spinning around, because the liquid carries on moving after you have stopped.

Young and old

People can usually hear sounds from about 20Hz, which is a low rumble, to about 18,000Hz, a high-pitched squeak. Children have the best hearing. They can hear very high-pitched sounds, higher than 20,000Hz, that older people cannot hear.

Too much noise

Listening to loud sounds, especially for a long time, damages your ears. People who work near noisy machinery wear ear-muffs to protect their ears.

The direction of sound

You know where a sound comes from because you have two ears. The ear closest to the sound hears it a little louder and slightly before the other one.

Loud and soft sounds

Some sounds are louder than others. Because sound spreads out in all directions, the further away you are from the sound, the quieter you hear it.

The loudness, or **intensity**, of a sound is measured in units called **decibels (dB)**. They are named after A.G. Bell who invented the telephone.

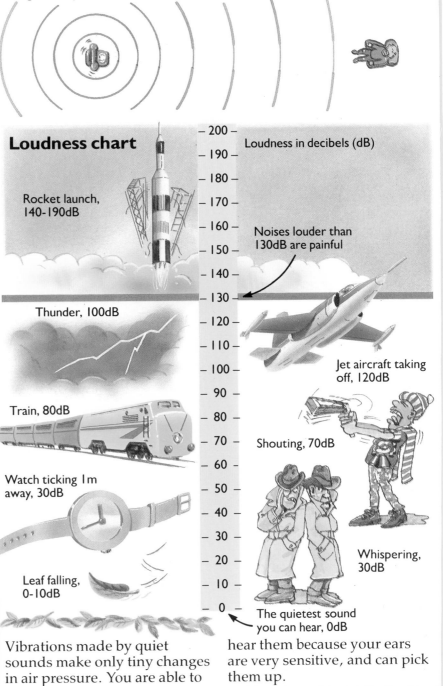

Loudness chart

Loudness in decibels (dB)

Rocket launch, 140-190dB

Noises louder than 130dB are painful

Thunder, 100dB

Jet aircraft taking off, 120dB

Train, 80dB

Shouting, 70dB

Watch ticking 1m away, 30dB

Whispering, 30dB

Leaf falling, 0-10dB

The quietest sound you can hear, 0dB

- 200 -
- 190 -
- 180 -
- 170 -
- 160 -
- 150 -
- 140 -
- 130 -
- 120 -
- 110 -
- 100 -
- 90 -
- 80 -
- 70 -
- 60 -
- 50 -
- 40 -
- 30 -
- 20 -
- 10 -
- 0 -

Vibrations made by quiet sounds make only tiny changes in air pressure. You are able to hear them because your ears are very sensitive, and can pick them up.

Musical sounds

All musical instruments work by making something vibrate. The sounds they make can be high or low, loud or quiet.

High and low

Low frequency

Long wavelength

High frequency

Short wavelength

The higher a sound, the higher its frequency*. This means that more vibrations reach you each second. So the distance between each vibration, called the **wavelength**, is smaller.

Loud and quiet

Loud sound

Big amplitude

Quiet sound

Small amplitude

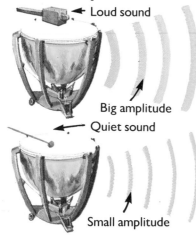

A loud sound makes big vibrations. The size of each vibration is called its **amplitude**. So the louder the sound, the bigger the amplitude.

How musical instruments work

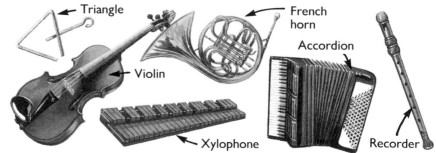

To make a higher sound, the string is shortened by pressing on it.

Pushing on a piano key makes a hammer hit the strings, so that they vibrate. Each note has two or three strings.

The sound of a double bass comes from making the strings vibrate by plucking them or by using a bow.

Blowing into a saxophone makes a thin piece of wood, called a **reed**, vibrate. This vibrates air in the saxophone.

The keys on a saxophone change the length of the column of air that vibrates inside it. The longer it is, the lower the sound.

Why do musical instruments sound different?

Triangle

French horn

Accordion

Violin

Xylophone

Recorder

Every musical instrument has its own special sound. Any note you hear has other high pitched notes, called **harmonics**, mixed into it. But they are too quiet for you to hear as separate notes. All instruments make different sounds because they have different harmonics from each other.

Hitting drums and cymbals with sticks or wire brushes makes them vibrate.

By making his lips vibrate as he blows, the player makes air in the trumpet vibrate.

Vibrating in time

You can play a piano without touching a key. Hold down the sustain pedal and sing a note. When you stop singing, you will hear the same note coming from the piano. The vibrations from your voice make the piano strings vibrate.

Open piano

The sustain pedal is the one on the right.

Pushing the sustain pedal leaves all the strings in the piano free to vibrate.

When the vibrations of one thing make something else vibrate, it is called **resonance**. Each string in a piano vibrates at one frequency, called its **natural frequency**. Your voice makes the string vibrate at its natural frequency.

Breaking glass

If you tap a glass, the sound you hear is made by the glass vibrating at its natural frequency. By singing loudly at this frequency, some singers can shatter a glass. Only a sound at the natural frequency of the glass can make vibrations big enough for this to happen.

Resonance makes sounds louder

Sound-box

Stringed instruments have a sound-box to make them louder. When the strings vibrate, the air inside the sound-box is made to vibrate by resonance.

Collapsing bridges

Everything has its own natural frequency. In 1940, the Tacoma Bridge, USA, collapsed. This was because wind made it vibrate at its natural frequency, causing huge vibrations. Soldiers do not march in step across bridges, in case their footsteps make the bridge vibrate at its natural frequency.

Seeing with sound

Night hunters

Some animals use sound to help them 'see'. Bats are able to find their prey at night and fly in the dark without bumping into things. Using sound to find things is called echo-location.

Bats send out lots of very high-pitched squeaks, then listen to the echoes bouncing off things. The shorter the time between the squeak and the echo, the closer they are to the object.

To help them catch moving insects, bats listen to the pitch* of the echo. The pitch changes as the insect moves past the bat. This is called the Doppler effect*.

Some moths can avoid being caught. They can hear the high-pitched sounds that bats make.

Bats can hear higher sounds than any other animal, up to 210,000Hz. The highest sounds people can hear are around 20,000Hz. Sound with a very high frequency is called **ultrasound**.

Sea sounds

Whales and dolphins also use echo-location to find their way through the sea. From the sound of the echo, they are able to tell what type of objects are around them.

Exploring underwater

Ships use ultrasound echoes to search for fish, to measure the depth of water beneath them, and to explore the ocean floor. This is called **sonar**. A computer can be used to build up a picture from the echoes.

Computer picture

High sounds spread out, or are diffracted*, much less than low sounds. That is why ultrasound is used for echo-location. It is so high that it hardly spreads out, so objects can be located very precisely from the echoes they make.

Looking for cracks

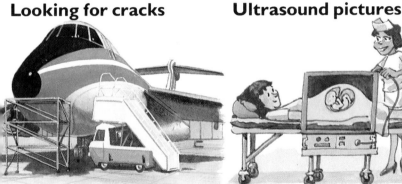

Ultrasound is used to test materials. Aircraft are checked this way. Engineers can tell from the echoes whether there are any cracks in the metal.

Ultrasound pictures

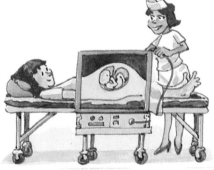

Ultrasound is used to look inside a mother at a growing baby. The echoes are changed into electrical signals and built up into a picture.

Exploring underground

Earthquakes or explosions cause huge vibrations through the Earth, called **seismic waves**. They travel at different speeds through different liquids and types of rock.

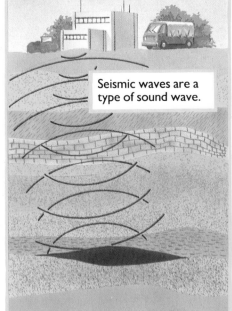

Seismic waves are a type of sound wave.

By measuring their speed, geologists can tell what the inside of the Earth is like. Seismic waves are also used to help people search for oil.

DID YOU KNOW?

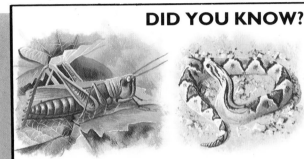

Not all animals hear sounds as you do. Grasshoppers 'hear' with their legs, waving them in the air to tell where a sound is coming from.

Snakes do not have ears so they cannot hear sounds through air. They pick up low sounds from the ground. Fish hear through their bodies.

Sounds from moving things

When a racing car passes by, its sound seems to change pitch. As it comes to you, the sound you hear gets higher. As it goes away, the sound gets lower. This is called the **Doppler effect**.

More vibrations reach you each second because the car is moving towards you, making the sound higher. As the car moves away, fewer vibrations reach you each second, making the sound lower.

What are things made of?

Look at all the different kinds of things around you. Everything you see is either a solid, a liquid or a gas. In this picture, you can see some of the differences between them.

You will also find lots of questions about the things around you. The next few pages answer these questions and explain what things are made of and how they change.

Solids keep their shape. They always take up the same amount of space.

Solids cannot be squashed into a smaller space.

Some solids are harder than others.

A solid cannot move unless something pushes or pulls it.

What makes things burn?

Some solids are light for their size, others are heavy for their size.

Some liquids are more difficult to pour than others.

Liquids cannot be squashed into a smaller space.

Some solids, like sand, are divided up into very small pieces.

Liquids do not have their own shape. They take the shape of whatever container they are in.

Liquids can move around and flow.

Both liquids and gases are able to flow. Both are sometimes called **fluids**.

A **vapor** is a gas given off by a liquid. You smell gasoline vapor when you are in a gas station.

What makes the wind blow?

Where does the rain come from?

Water can be a liquid, a solid or a gas. When water freezes, it turns into a solid, ice. When water boils, it turns into a gas, steam.

Why does sugar dissolve in coffee?

How does a thermometer work?

Why does ice-cream melt when it gets warm?

How do smells reach you?

What makes a drink fizzy?

Gases can be squashed.into a smaller space.

Why can you dry yourself with a towel?

Where does a puddle go when it dries up?

Gases can move around and flow. They have no shape of their own. They spread out and fill the container they are in, and take its shape.

Air is a mixture of different gases. Most gases are invisible.

DID YOU KNOW?

Diamond is the hardest thing known in the world. It is so hard that it can even cut through glass. Man-made diamonds are used in many cutting and drilling machines.

Atoms and molecules

Everything around you is made up of tiny pieces, called atoms and molecules, that are much too small to see. Atoms are so small that over 100,000 million would fit in a full-stop.

A grain of sand contains 50,000 million million molecules. Each molecule is made up of three atoms.

Silicon atom

Oxygen atom

Imagine you could divide a grain of sand into smaller and smaller pieces. You would eventually get a piece, or **particle**, that could not be divided up any more. This is called a molecule, the smallest possible piece of sand.

A molecule of hydrogen gas contains two identical atoms.

Hydrogen

An oxygen molecule contains two identical atoms.

Oxygen

Everything in the Universe is made of different atoms and molecules. Molecules are made of two or more atoms joined together. Most molecules contain a few atoms, but others may contain thousands of atoms.

A water molecule contains three atoms, two hydrogen atoms and an oxygen atom.

Oxygen

Hydrogen

What's inside an atom?

There are about 105 different types of atoms known. They are made up of even smaller particles, called **protons**, **neutrons** and **electrons**.

The different atoms contain different numbers of protons, neutrons and electrons. This picture shows what the inside of an atom may look like.

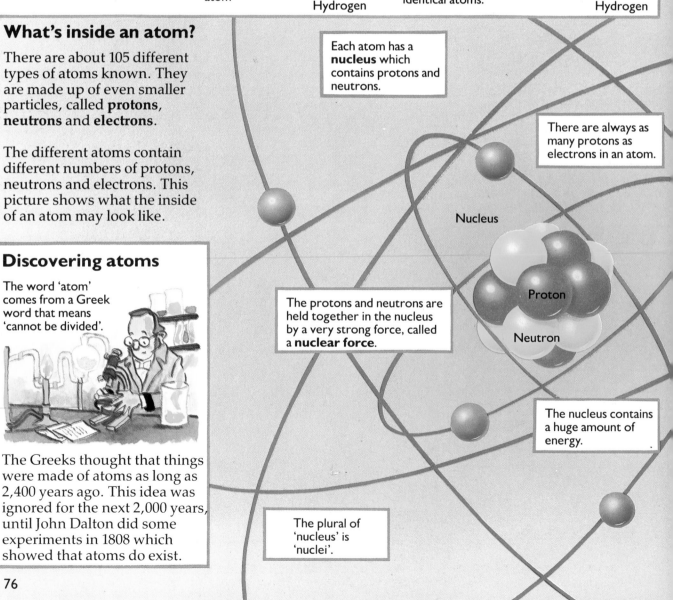

Each atom has a **nucleus** which contains protons and neutrons.

There are always as many protons as electrons in an atom.

Nucleus

Proton

Neutron

The protons and neutrons are held together in the nucleus by a very strong force, called a **nuclear force**.

The nucleus contains a huge amount of energy.

The plural of 'nucleus' is 'nuclei'.

Discovering atoms

The word 'atom' comes from a Greek word that means 'cannot be divided'.

The Greeks thought that things were made of atoms as long as 2,400 years ago. This idea was ignored for the next 2,000 years, until John Dalton did some experiments in 1808 which showed that atoms do exist.

Whizzing electrons

Electrons whiz around the nucleus. They are held in the atom by an electrical force. They have an **electric charge** which means that they carry electricity. There are two types of electric charge, **positive charge** and **negative charge**.

Electrons have a negative charge, protons have a positive charge, neutrons have no charge at all. Because there are the same number of electrons and protons, the positive charges and the negative charges in an atom are balanced equally.

Tiny electrons move around the nucleus. They are very light. One electron weighs about 1/2000 of a proton.

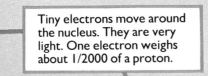

Electron

The inside of an atom is mostly empty space. The nucleus is about 10,000 times smaller than the atom itself.

DID YOU KNOW?

Atoms and molecules are so tiny that there are about as many atoms in a grain of sand as there are grains of sand on a beach.

Molecules move

Put a few drops of ink into a glass of water. Eventually, all the ink will spread evenly through the water. This is because the molecules in liquids are always moving around and bumping into each other.

In the same way, molecules in gases are always moving around in all directions. This is why you can smell flowers across a room. Their scent reaches you because their molecules spread through the air.

When molecules spread through liquids and gases, this is called **diffusion**. Gas molecules move much faster than liquid molecules. So it takes longer for ink to spread through water than for smells to spread through air.

Nuclear energy

When a nucleus is split in two or when two nuclei are joined to form a new nucleus, a huge amount of energy, called **nuclear energy**, is given out. Splitting a nucleus is called **fission**, joining up nuclei is called **fusion**.

When nuclear energy is let out slowly, it can be used to produce electricity in nuclear power stations. But if it is let out all at once, it makes an enormous explosion. This is how a nuclear bomb works.

Nuclear power stations use **uranium** fuel. The place where the uranium nuclei are split up is called the **reactor**. It is covered in thick concrete to stop deadly **nuclear radiation** leaking out.

Solids, liquids and gases

Why can you push your finger through jam, but not through steel? Why can you pour water, and not wood? Why does salt mix into water and disappear, but sand stay separate? What makes solids, liquids and gases different from each other?

Solids

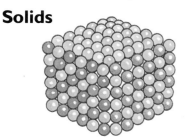

The atoms in solids are held very close together. They vibrate all the time, but, because they are held in rows by very strong forces, they cannot move around each other.

Solids cannot be squashed into a smaller space, as their atoms are already so close together. They keep their shape because their atoms are held together by such strong forces.

Liquids

The molecules in liquids are close to each other, but the forces holding them together are not as strong as in solids. The molecules can move around

Liquids take the shape of their container.

and change places with each other, so liquids can flow and change shape. Liquids cannot be squashed because their molecules are so close together.

Gases

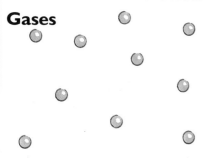

In gases, the molecules are always moving around very quickly in all directions. The forces holding the molecules together are very weak, so

Gases do not have a fixed shape. They spread out to fill their container.

gases are able to spread out and flow. Because there are such big gaps between the molecules, gases can be easily squashed into a smaller space.

Liquids hold together

A drop of liquid seems to be held together by an elastic skin. This happens because of surface tension*. Drops form because the liquid molecules are pulled towards each other.

The hairs on a paintbrush cling together when they are wet, but not when they are dry. This is because the water molecules are pulled towards each other.

A liquid spreads over a surface if the pull of the molecules from the surface is stronger than the forces holding the liquid molecules together.

The forces between the water molecules keep them held together in drops.

Water rolls off ducks' feathers because they are covered in a layer of grease which does not pull the water molecules to it.

How do things soak up water?

A towel will soak up water. The water is sucked into tiny spaces between the threads of the towel. Sucking liquids up this way is called **capillary action**. Plastic does not soak up liquids because it has no holes in it.

Water creeps up to the leaves of plants from the soil. This is because their roots contain tiny tubes which suck up water by capillary action.

Viscosity

Some liquids, like water, are runny and easy to pour. Others, like honey, are thicker and pour more slowly. How thick a liquid is, is called its **viscosity**.

Try pouring clear honey after leaving it in the fridge for a few hours. Liquids become thicker, or more viscous, the colder they get. As they warm up, they become less viscous.

DID YOU KNOW?

Glass is not a solid, but a liquid. You cannot see it flowing because it is very viscous. Very old windows are thicker at the bottom because the glass has flowed downwards over the years.

Solutions

When you add sugar to coffee, the sugar and coffee mix together and form a **solution**. The sugar **dissolves** in the coffee.

The thing that is dissolved, like salt, is called a **solute**. The thing that does the dissolving, like water, is called a **solvent**.

The sugar molecules spread out evenly through the coffee.

If the coffee is hot, and if you stir it, the sugar dissolves more quickly.

A gas called carbon dioxide is dissolved into drinks to make them fizzy.

Oil

Vinegar

Oil does not dissolve in vinegar. This is why oil and vinegar salad dressing separates into two layers.

The sea is a solution of salt in water.

Salt dissolves in water, but sand does not.

Some things will not dissolve in others. Oil and grease do not dissolve in water. Dry cleaning can get oil stains out of clothes by using another solvent, a chemical called **tetrachloromethane**. It is only used in dry cleaning shops because it gives off poisonous fumes.

Heating and cooling

What happens when things get hot?

As things heat up, they become slightly larger. They occupy more space than when they are cold. This is called thermal expansion. As they cool down again, they shrink back to their normal size. This is called contraction.

Gases expand

Put an open, plastic bottle in a fridge. When it is cold, put a balloon over its top. Then put the bottle in a bowl of hot water. Watch the balloon blow up by itself.

This happens because the air **expands** as it is heated. Put the bottle back in the fridge. The balloon will go down again, because the air shrinks, or **contracts**, as it cools down.

Liquids expand

A thermometer works by thermal expansion. It contains liquid in a glass tube. As the thermometer is heated, the liquid rises up the tube. This is because the liquid expands more than the glass. As the thermometer cools, the liquid shrinks back down again.

Solids expand

A 100m steel rail may get 4cm longer on a hot day.

Solids also expand when they get hotter. The central part of the Humber Bridge, England is 1,410m long. During the summer it can become over half a metre longer than it is in winter.

Allowing for expansion

How much something expands depends on its size, what it is made of, and how much it is heated. Small things expand by small amounts. Large things expand by large amounts.

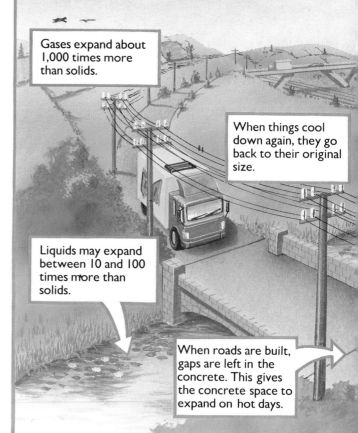

Gases expand about 1,000 times more than solids.

When things cool down again, they go back to their original size.

Liquids may expand between 10 and 100 times more than solids.

When roads are built, gaps are left in the concrete. This gives the concrete space to expand on hot days.

Why things expand when they get hot

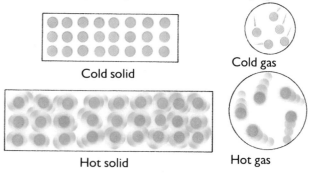

Cold solid

Cold gas

Hot solid

Hot gas

Atoms and molecules are moving and bumping into each other all the time. The hotter they get, the faster they move, and the harder they hit each other. This makes them take up more space.

Gases expand more than liquids and liquids expand more than solids. The bigger the temperature change, the bigger the change in size.

Railway lines are built so they can expand on hot days.

Steel expands more than wood.

Telephone wires are hung slackly. This is so that they do not snap apart when they contract on cold winter days.

Using expansion

This only works if there is no hole in the ball.

If the metal lid on a jar is stuck, hold it under a hot water tap. Because metal expands more than glass, the lid will loosen.

If a ping-pong ball is squashed, put it in warm water. The air inside it expands, pushing the ball back into shape again.

Expansion and density

How heavy something is for its size, its density*, depends on how close its atoms are to each other. When something expands, its atoms move further apart, so it becomes less dense.

Warm air rises. This is because, as it expands and becomes less dense, it floats up through colder air. Heat is carried through liquids and gases this way. This is called convection*.

DID YOU KNOW?

Guitars often have to be tuned on stage because the strong lights heat them up. Steel expands more than wood, so the steel strings become slack and out of tune.

Expansion can be dangerous

Never pour boiling water into a glass. Glass conducts heat badly, so the inside gets hot and expands. But the outside stays cold and does not expand. This makes the glass shatter.

You should never let an aerosol heat up or throw it in a fire. This is because there are gases sealed inside aerosols. If heated, the gases would expand, making the can explode.

Boiling and freezing

Things can be solids, liquids or gases, and they can change from one to another. If you freeze water, it turns to ice. If you boil water, it turns to steam. When you melt ice, it changes from a solid back into a liquid, water.

When you boil water, it turns into a gas, steam. To change something from a solid to a liquid and then from a liquid to a gas, you have to heat it. This gives it heat energy. This added energy makes the molecules move around faster.

Gas

Water boils and turns into steam at 100°C. This is called its **boiling point**.

As water is boiling, its temperature stops going up. It stays steady at 100°C.

As steam cools below 100°C, it turns back into water. This is called **condensation**.

Liquid

The gas from water is called **steam** or **water vapor**.

Liquid

Ice melts into water at 0°C. This temperature is called its **melting point**.

Water freezes and turns into ice at 0°C. This is called its **freezing point**.

Solid

As a solid is heated, the molecules make bigger and bigger vibrations. Eventually, the molecules cannot be held in their fixed positions any more. When this happens, the solid melts and turns into a liquid.

As a liquid is heated, the molecules move around faster and faster until they fly off, forming a gas. When a liquid gets hot enough, it boils. Gas bubbles form in the liquid and rise to the surface.

To change a gas into a liquid, or a liquid into a solid, you have to cool it. This takes out heat energy and slows down the molecules. To turn water into ice, you have to cool it in a fridge to take out the energy.

Why do wet things get dry?

Liquids are evaporating and changing into vapor all the time.

Sweat cools you down as it evaporates from your skin.

Where does the water go when a puddle dries up? The water slowly changes into vapor and spreads out through the air. This is called **evaporation**.

Your skin feels cold when it is wet. This is because water is evaporating from it. The water takes heat energy from your skin as it changes into vapor.

Faster evaporation

Water evaporates more quickly from wet clothes when it is hot, when the wind is blowing, and if they are spread out so that more air can get to them.

Boiling and freezing temperatures

Different things boil and freeze at different temperatures. Water freezes at 0°C and boils at 100°C. Steel melts at over 1,400°C. Cooking oil boils at over 200°C.

Changing melting and boiling points

The melting point and the boiling point of water changes if you mix salt in with it. Salty water freezes at a lower temperature and boils at a higher temperature than pure water.

Mercury freezes at −39°C. In very cold places, mercury thermometers cannot be used.

Pure water freezes at 0°C, but water with salt mixed into it freezes at −20°C.

Putting salt on roads in winter stops water freezing on them so they do not get icy.

Burst pipe

Food cooks faster in salty water, because salty water boils at a higher temperature than pure water.

When most liquids freeze into solids, they take up less space. Water is unusual because, when it freezes into ice, it takes up *more* space. This is why water pipes can burst in winter. The water inside them freezes, splitting the pipe open.

When any liquid changes into gas, the gas takes up much more space. This is what makes a steam engine* work. Water is boiled into steam inside the engine. The steam takes up much more space than the water, pushing the pistons in and out.

Cool drinks

Foggy windows

DID YOU KNOW?

At the top of Mount Everest, where the air pressure is lower, water boils at only 70°C. The lower the air pressure, the lower the boiling point of a liquid.

Ice-cubes cool a drink as they melt. Like all solids, ice needs heat energy to melt. It takes this energy from the liquid around it, cooling the drink.

When water vapor from your breath hits a cold window, it turns into tiny drops of water, making the window fog up. This is called **condensation**.

*Steam engine, 44.

The weather

Where does rain come from? What is snow? Why does the weather change? What makes the wind blow? All the world's weather is caused by the Sun, Earth, air and water.

What makes it rain?

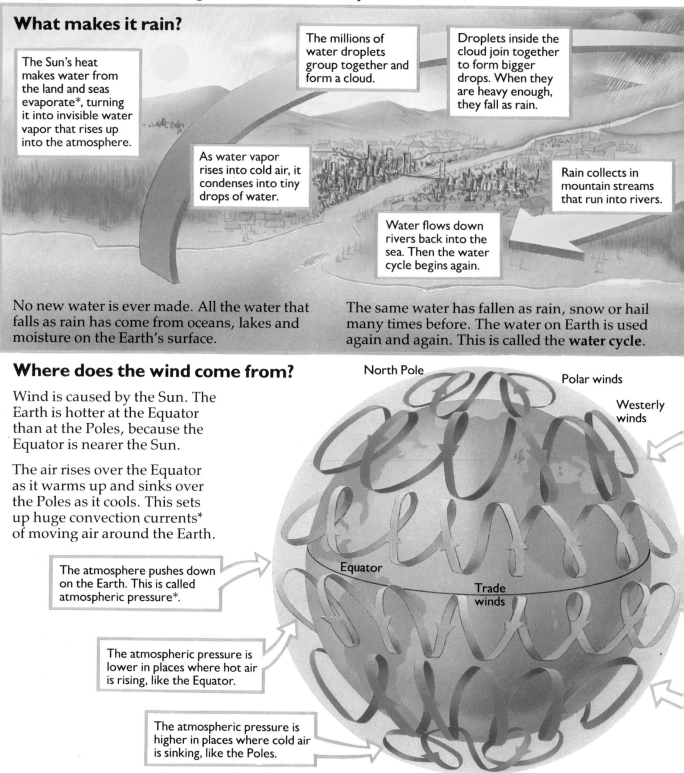

The Sun's heat makes water from the land and seas evaporate*, turning it into invisible water vapor that rises up into the atmosphere.

The millions of water droplets group together and form a cloud.

Droplets inside the cloud join together to form bigger drops. When they are heavy enough, they fall as rain.

As water vapor rises into cold air, it condenses into tiny drops of water.

Rain collects in mountain streams that run into rivers.

Water flows down rivers back into the sea. Then the water cycle begins again.

No new water is ever made. All the water that falls as rain has come from oceans, lakes and moisture on the Earth's surface.

The same water has fallen as rain, snow or hail many times before. The water on Earth is used again and again. This is called the **water cycle**.

Where does the wind come from?

Wind is caused by the Sun. The Earth is hotter at the Equator than at the Poles, because the Equator is nearer the Sun.

The air rises over the Equator as it warms up and sinks over the Poles as it cools. This sets up huge convection currents* of moving air around the Earth.

The atmosphere pushes down on the Earth. This is called atmospheric pressure*.

The atmospheric pressure is lower in places where hot air is rising, like the Equator.

The atmospheric pressure is higher in places where cold air is sinking, like the Poles.

North Pole

Polar winds

Westerly winds

Equator

Trade winds

South Pole

*Atmospheric pressure, 41; Convection currents, 16; Evaporation, 82.

Not all clouds make rain. If a cloud moves into warmer air, it evaporates.

If a cloud reaches very cold air, the water droplets turn into ice crystals.

If the ice crystals do not melt, they fall as hailstones or snowflakes.

Crystals

Snowflakes are made of tiny, regularly shaped pieces of ice. These are called **crystals**. No two snowflakes are the same. Each one contains ice crystals of different shapes and sizes.

As areas of high and low atmospheric pressure move around the world, the weather changes from day to day.

The wind is caused by air flowing from areas of high pressure to areas of low pressure.

The winds are swung sideways as the Earth spins.

Each half of the Earth has three main wind patterns. These are the **trade winds**, the **polar winds** and the **westerlies**.

Humidity

The amount of water vapor in the air is called **humidity**. Warm air can carry more water vapor than cold air. When it is very humid, your skin feels sticky because there is so much water in the air already that your sweat cannot evaporate.

Dew and frost

After a cold night, you can see **dew** outside. This is because cold night air holds less water vapor than hot air, so the water vapor condenses into drops of water. If the temperature is less than 0°C, the water vapor freezes and forms **frost**.

DID YOU KNOW?

The largest hailstone ever found was in Kansas, USA. It was 19cm wide and weighed 758g, the size of a melon.

Great storms

Huge storms of whirling wind and rain form over seas near the Equator where the air is warm and damp. They are called **hurricanes**, **typhoons** or **cyclones** depending on where they form.

A **tornado** is a funnel-shaped storm that is usually about 100m across. The hot air at its center whirls at over 600kph, and sucks up everything in its path, causing terrible damage.

Elements and compounds

All the different things in the Universe are made from atoms. There are about 105 different kinds of atom known today. Everything around you is made of these atoms combined in different ways.

Things that are made of one kind of atom are called elements. Because there are 105 kinds of atom, there are 105 different elements. Things that are made of different kinds of atoms joined together are called compounds.

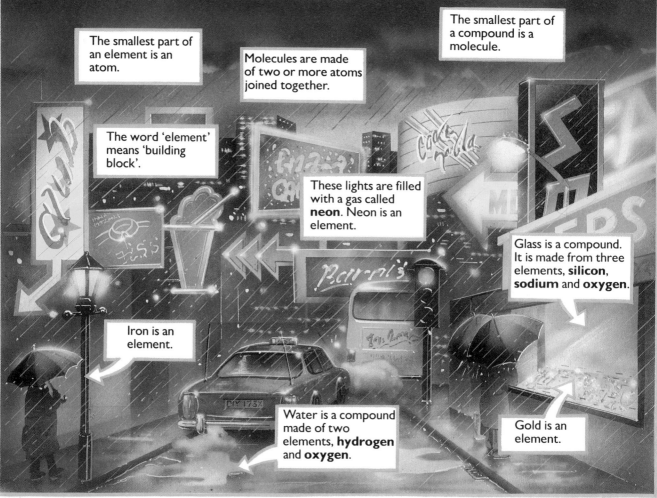

The smallest part of an element is an atom.

The smallest part of a compound is a molecule.

Molecules are made of two or more atoms joined together.

The word 'element' means 'building block'.

These lights are filled with a gas called **neon**. Neon is an element.

Glass is a compound. It is made from three elements, **silicon**, **sodium** and **oxygen**.

Iron is an element.

Gold is an element.

Water is a compound made of two elements, **hydrogen** and **oxygen**.

Naming elements and compounds

The element hydrogen has the symbol H.

The element oxygen has the symbol O.

The symbol for water is H_2O.

Each element has a symbol, which may be one or two letters. The same symbols are used for compounds. They show which elements the compound contains. A number after the symbol shows how many atoms of each element there are in a molecule. Water is a compound. Each water molecule contains two atoms of hydrogen and one atom of oxygen. Its symbol is H_2O.

Making something new

Salt is a compound called **sodium chloride**.

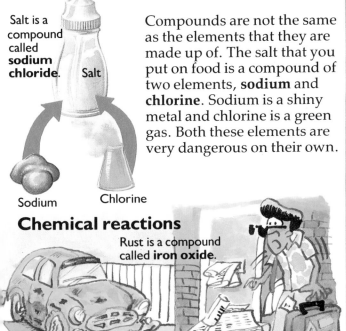

Sodium   Chlorine

Compounds are not the same as the elements that they are made up of. The salt that you put on food is a compound of two elements, **sodium** and **chlorine**. Sodium is a shiny metal and chlorine is a green gas. Both these elements are very dangerous on their own.

Chemical reactions

Rust is a compound called **iron oxide**.

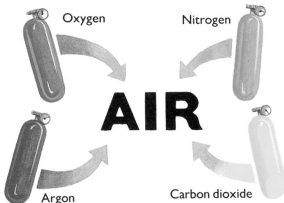

Anything made of iron gets rusty if you leave it outside for a long time. Atoms of iron join up with oxygen atoms in air, making a new compound, **rust**. This is called a **chemical reaction**. A compound is made whenever atoms of different elements join, or **react**, together.

Air is a mixture

Oxygen   Nitrogen

AIR

Argon   Carbon dioxide

Air contains several gases. The atoms of these gases are mixed up, but not joined together. So air is called a **mixture**, *not* a compound. It contains three elements, **nitrogen**, **oxygen** and **argon**, and one compound, **carbon dioxide**.

Alchemy

For many hundreds of years, people thought that all things on Earth were made from **air**, **earth**, **fire** and **water**. By mixing them in different amounts, they thought they could turn one thing into another. Some people, called **alchemists**, tried to change ordinary metals into gold.

The Periodic Table

About 100 years ago, a Russian scientist called Dmitry Mendeleyev listed all the known elements in a chart called the **Periodic Table**. This chart groups elements that are similar and shows which ones will react together to form compounds.

DID YOU KNOW?

Diamond and pencil lead, which is called **graphite**, are made from the same thing. Both of them are forms of the element, **carbon**. They are different because the carbon atoms inside them are held together in different ways.

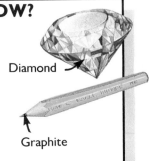

Diamond

Graphite

Fire

If something gets hot enough, it burns. Once it starts burning, it gives out so much heat energy that it carries on burning on its own.

People rely on burning fuels for cooking, heating and for working machines. But when fire gets out of control, it is very dangerous.

What happens when things burn?

Burning is a chemical reaction*. Things burn when they get hot enough to react with the oxygen in the air around them.

As in all chemical reactions, burning produces new compounds. Smoke and ash are a mixture of these compounds.

Fire needs three things: heat, fuel and oxygen. If any one of these things is taken away, the fire goes out.

Oxygen

Heat

Fire

Fuel

What makes soot?

Soot is a powder made of tiny pieces of the element **carbon**. Many things, like wood and coal, contain carbon. When they burn the carbon reacts with oxygen, producing fumes. But if there is not enough oxygen to react with the carbon, soot is produced.

Dangerous fumes

Fire uses up oxygen in the air and produces fumes that often are as dangerous as the flames. The fumes from plastics, foam

rubber and some paints are deadly, even in small amounts. This is why fire-fighters carry air cylinders and wear masks.

How fires spread

Fire can spread by convection*. Convection currents carry heat, smoke and burning materials to other places. Where they land, they can start new fires.

Fire can spread by radiation*. Heat radiation from the flames heats things that are close to the fire. Eventually they get so hot that they burst into flames.

Fire can spread by conduction*. Even though metal does not burn, it can carry the heat from a fire by conduction, setting other things alight.

Putting out fires

If you discover a fire, shout "Fire!" to warn people. Once you are safe, quickly telephone the fire department. Never try to tackle a fire on your own.

Taking away either the fuel, heat or oxygen from a fire will put it out. Depending on what is burning, fires have to be put out in different ways.

Shutting doors and windows cuts out the supply of fresh oxygen. This slows down the spread of the fire.

Spraying water on a wood or paper fire takes the heat away. Without heat, the fire goes out.

If someone's clothes are on fire, rolling them on the ground in a rug or curtain cuts out the oxygen.

Putting a lid or a wet towel over a pan of burning fat or oil cuts out the oxygen. *Never* pour water on burning oil. The oil will splash and carry on burning as it floats on the water.

If something electric is on fire, the power should be switched off. Gas or powder fire extinguishers can be used to put the fire out, but *never* use water because it conducts electricity*.

Fire extinguishers

Fire extinguishers are filled with water, foam, powder or a gas. They are used for putting out different kinds of fires.

Water is used for most fires *except* burning liquids and electrical fires.

Foam is used for putting out burning liquids. It should *never* be used on electrical fires.

Dry powder is used for putting out burning liquids and electrical fires.

Carbon dioxide is used for putting out burning liquids and electrical fires.

Halon gas is used for putting out burning liquids and electrical fires.

Never breathe in the fumes given out by halon or carbon dioxide extinguishers.

Burning and engines

Burning produces hot gases that take up much more space than the thing that burns. The hot gases from burning fuel make car engines work.

As the gases expand in the engine, they make the pistons* move in and out. Rocket and jet engines are pushed forward as the hot gases rush out behind.

*Electricity, 96; Pistons, 45. **89**

Materials

The things around you are all made out of different materials. Some materials come from plants, animals or things that are dug up from the ground. These are sometimes called natural materials. Others are made in factories. These are called man-made or synthetic materials. Scientists study atoms and molecules so they can design new synthetic materials.

Metals

Many things are made out of metals, or mixtures of different metals, called **alloys**. Metals are found in **ores** that are dug out of the ground. The metals are removed by heating up the ores.

Copper and **bronze** were the first metals that people used. Today, **iron** and **steel** are the most widely used metals. Steel is a mixture of iron and a little carbon.

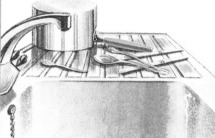

Stainless steel is a special type of steel that does not rust*. **Aluminum** is a very light metal, so is used to make parts for aircraft.

Ceramics

For thousands of years, people have used **clay** to make pots and jugs. The clay is shaped when damp, and then baked in an oven to harden it. Materials like clay are called **ceramics**.

Ceramic materials are used for many things. **Porcelain** is used to make cups and plates. **Bricks** and **tiles** are used for building. Even **glass** is a type of ceramic.

Ceramics can be heated to very high temperatures. New, strong ceramic materials are now being used to make parts for engines.

Fibers

Cloth and fabric are made from thin strands, called **fibers**. Some fibers, like wool, silk and cotton, come from plants and animals. They are called **natural fibers**.

Many fabrics are made from man-made fibers like nylon, rayon and polyester. A special fiber, **kevlar**, is even stronger than steel but very light. It is used in some aircraft and boats.

Plastics

Most plastics are made from compounds* that are found in crude oil*. There are many different types of plastics that are used for different things.

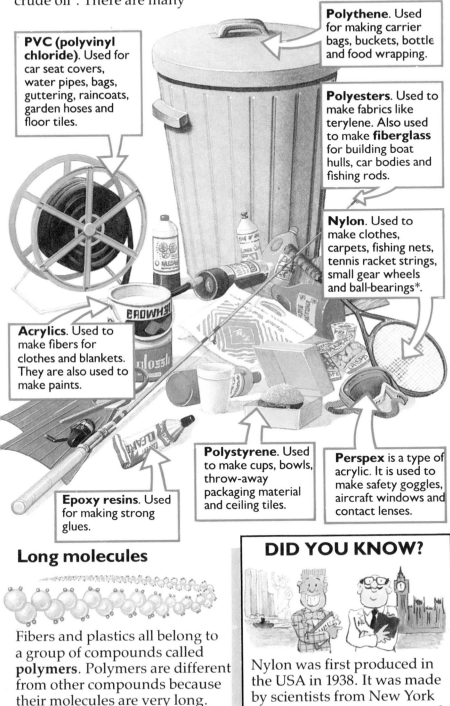

PVC (polyvinyl chloride). Used for car seat covers, water pipes, bags, guttering, raincoats, garden hoses and floor tiles.

Polythene. Used for making carrier bags, buckets, bottle and food wrapping.

Polyesters. Used to make fabrics like terylene. Also used to make **fiberglass** for building boat hulls, car bodies and fishing rods.

Nylon. Used to make clothes, carpets, fishing nets, tennis racket strings, small gear wheels and ball-bearings*.

Acrylics. Used to make fibers for clothes and blankets. They are also used to make paints.

Epoxy resins. Used for making strong glues.

Polystyrene. Used to make cups, bowls, throw-away packaging material and ceiling tiles.

Perspex is a type of acrylic. It is used to make safety goggles, aircraft windows and contact lenses.

Long molecules

Fibers and plastics all belong to a group of compounds called **polymers**. Polymers are different from other compounds because their molecules are very long. They are made of lots of small molecules joined together.

DID YOU KNOW?

Nylon was first produced in the USA in 1938. It was made by scientists from New York and London, and was named after these cities (NY-Lon).

Using materials with care

Trees are cut down to provide wood, but this destroys whole forests that can never be replaced. These forests are needed to balance gases* and moisture* in the atmosphere.

Natural materials, like wood, are called **biodegradable** because they rot when they are thrown away. Most man-made materials, like plastics, cause pollution because they never rot away.

New plastics are now being made that *are* biodegradable. These materials are not made from crude oil. At the moment, they are more expensive to make, but they do not cause pollution.

---

*Balancing gases, 21; Ball-bearings, 31; Compounds, 86; Crude oil, 25; Moisture in the air, 84.

Electricity around you

Think how often you watch television, switch on lights and use telephones. These things, and many others, work using electricity. A world without electricity would be very different.

In this picture, you will find lots of questions about things that work because of electricity. You will find the answers to these questions in the next few pages.

Electricity was not invented. It was first discovered by the Greeks, about 2,000 years ago. But people only learned how to produce it and make use of it just over 150 years ago.

Electricity is a form of energy. It can be changed into heat energy, light energy and sound energy. It can also be changed into kinetic energy to make machines work.

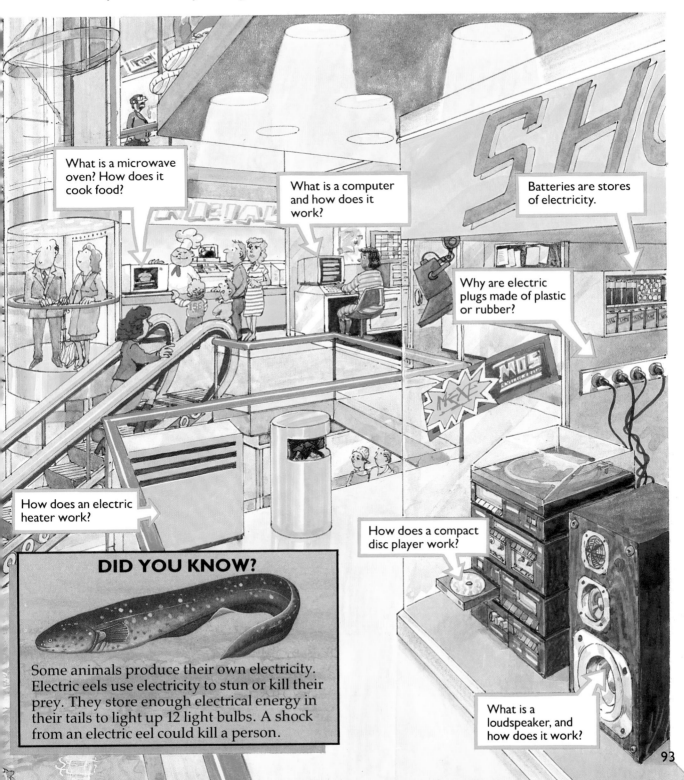

What is a microwave oven? How does it cook food?

What is a computer and how does it work?

Batteries are stores of electricity.

Why are electric plugs made of plastic or rubber?

How does an electric heater work?

How does a compact disc player work?

DID YOU KNOW?

Some animals produce their own electricity. Electric eels use electricity to stun or kill their prey. They store enough electrical energy in their tails to light up 12 light bulbs. A shock from an electric eel could kill a person.

What is a loudspeaker, and how does it work?

Flowing electricity

Many things around you need electricity to make them work. Some things, like flashlights, use electricity from batteries.

Other things, like lamps and televisions, are plugged in. They use mains electricity that is produced in power stations*.

Conductors and insulators

Electricity can flow, or is **conducted**, more easily through some materials than through others. Things through which electricity flows easily are called **conductors**. Things that electricity cannot flow through are called **insulators**.

Glass and other ceramics are insulators.

Air is an insulator.

Wood is an insulator.

Inside this plastic cable there are metal wires that carry electricity.

Rubber is an insulator.

Water conducts electricity.

The prongs on a plug are made of metal to conduct electricity from the socket.

Plugs are made of rubber or plastic because they are insulators.

Metals are good conductors. This is why metals are used to make wires for carrying electricity.

Most plastics are insulators. So metal wires are covered in plastic to stop you getting an electric shock.

What is electricity?

The electrons in atoms carry an electric charge*. When electrons flow together in one direction, they carry electricity with them. Flowing electricity is called **current electricity**.

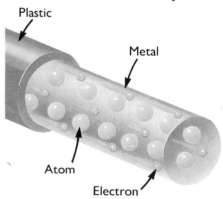

Plastic

Metal

Atom

Electron

Things that conduct electricity, like metals, have electrons* that are free to move. This is because the electrons are not held tightly to their atoms*. The electrons are able to carry electricity from place to place.

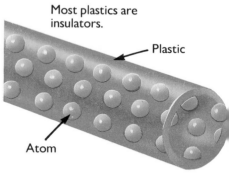

Most plastics are insulators.

Plastic

Atom

The electrons in insulators are held tightly inside their atoms. Because the electrons cannot move, insulators cannot conduct electricity.

The amount of electricity that flows through a wire each second is called an **electric current**. It is measured in **amperes** or **amps (A)**.

Resistance

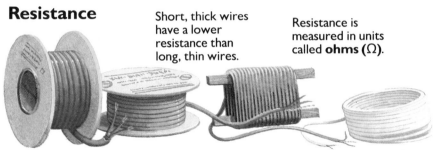

Short, thick wires have a lower resistance than long, thin wires.

Resistance is measured in units called **ohms (Ω)**.

Electricity flows better through some things than others. How well something conducts electricity is measured by its **resistance**. The resistance of a wire depends on what it is made of, its length and its thickness.

The *lower* the resistance of a wire, the *better* it conducts electricity. Copper is used to make wires because it it has a lower resistance than most other metals and therefore conducts electricity better.

Electric circuits

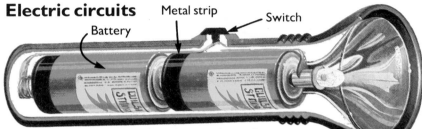

Battery — Metal strip — Switch

The batteries and the metal strip form an electric circuit.

An electric current will only flow around a continuous wire. This is called a **circuit**. The current stops flowing if there are any breaks in the circuit.

An electric current can be turned on and off with a **switch**. Switching something on joins up the circuit. Switching something off breaks the circuit.

Electricity into heat

When electricity flows, it makes things heat up. The higher the resistance of a wire, the hotter it gets when electricity flows through it. This is why the coiled up wires inside a hair-drier glow red hot.

Electricity into light

A light bulb contains a thin wire. As electricity flows through the wire, it glows white hot, giving out light. Only 2% of the electrical energy going into a light bulb changes into light. The rest changes into heat.

Batteries

A battery contains a store of chemical energy. This energy changes into electrical energy when the battery is connected in a circuit.

Batteries provide the electric force that pushes electrons around a circuit. This force is called an **electromotive force** and is measured in **volts (V)**.

Types of electricity

Electricity from power stations is called mains electricity. It is much more powerful than the electricity from batteries.

Circuit breakers and fuses

Things are damaged if too much electric current flows through them. Circuit breakers cut the electricity if the current gets too large. Houses have circuit breakers to protect the wiring.

The most common type of circuit breaker is called a **fuse**. It is a piece of special wire that melts if too much current flows through it, breaking the circuit.

When you plug something in and turn it on, mains electricity flows into it from the socket. Each plug has a fuse inside it, to break the circuit if too much current flows through it.

Electric current from batteries flows around circuits in one direction. This is called **direct current**. Mains electricity is different. The current changes direction many times each second. This is called **alternating current**.

Electricity cables contain two wires called **live** and **neutral**. Both carry live electric current. Some cables have a third wire called an **earth** wire. If a circuit is faulty, the earth wire conducts the electricity safely into the ground.

Safe electricity

Mains electricity can be very dangerous. *Never* touch anything with live electricity flowing through it, because you could get a deadly electric shock.

Never use anything with a cable that has the plastic insulation worn through. Touching the cable could give you a shock.

Never plug too many things into one socket. This can make too much current flow through the socket and can cause a fire.

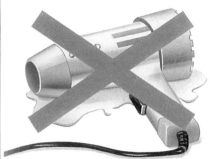

Never use anything that works off mains electricity when you are wet. This is because water conducts electricity very well. This is why water must never be sprayed on an electrical fire*.

*Electrical fires, 89.

Static electricity

Current electricity can flow. But there is another type, **called static electricity, that stays in one place.**

Rub a balloon on a wool sweater and hold it up to a wall. It will stick there by itself. Now rub two balloons on the sweater and put them beside each other.

They will move away from each other without your touching them. These things happen because rubbing the balloons gives them static electricity.

Atoms* contain electrons that carry negative charge and protons that carry positive charge. Normally, there are the same number of protons and electrons in an atom so the positive and negative charges cancel each other out. But when you rub the balloon, it picks up some extra electrons from the wool and becomes electrically charged.

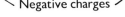

Negative charge

Positive charge

Negative charge

Negative charges

The extra negative charges in the balloon are attracted to the positive charges in the wall, so it sticks there. Negative charges are always attracted to positive charges.

The two balloons push away from each other because both have extra negative charges. Negative charges always repel negative charges and positive charges repel positive charges.

Static around you

Rubbing your shoes on a nylon carpet makes static electricity build up on you. If you touch something metal, you may feel a tiny shock as a spark jumps from you to the metal.

Magnets and electricity

Magnets can be used for many things. If you spill pins on the floor, a magnet will help you pick them up by pulling the pins to it. A compass, which helps people find their way, has a magnet inside it. Many things that work using electricity have a magnet inside them. Magnets are used to make electric motors spin and to produce electricity in generators.

The magnet picks up steel pins because steel contains iron.

See what you can pick up with a magnet. You cannot pick up anything made of plastic, wood or rubber. But things made of the metals iron, cobalt or nickel are pulled towards it.

Things will be pulled to a magnet if they are inside its magnetic field. They do not need to touch the magnet.

North pole

South pole

Magnets produce a **magnetic force**. The area around a magnet where the force works is called a **magnetic field**. It is strongest at the ends of the magnet, which are called **poles**.

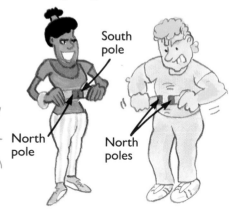

South pole

North pole

North poles

Hold two magnets together. Feel how they pull and push each other as you turn them around. This happens because unlike poles are pulled together, but like poles are pushed apart.

Magnets and compasses

The Greeks used magnets 2,000 years ago. They dug a material out of the ground, called **lodestone**, which is magnetic. They found that if a magnet is able to swing freely, its north pole always points North and its south pole points South. This happens because the Earth has its own magnetic field.

The Earth's magnetic field is strongest at the Poles.

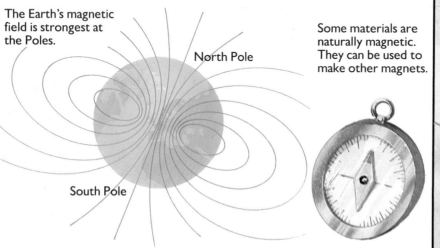

North Pole

South Pole

Some materials are naturally magnetic. They can be used to make other magnets.

The needle inside a compass is a magnet. The magnet points North, so you can tell which direction you are facing.

Sailors have used compasses to help them find their way at sea, or **navigate**, since the 11th century.

Electromagnetism

Electricity and magnetism together make many things work. This is called **electromagnetism**. Electricity can be used to make a magnetic field and magnetism can be used to produce electricity. Whenever an electric current flows through a wire, it produces a magnetic field around the wire. When the electric current is switched off, the magnetic field disappears.

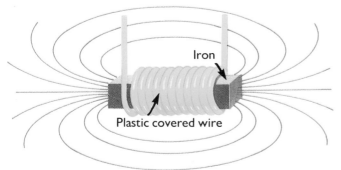

Iron

Plastic covered wire

The bigger the electric current, the stronger the magnetic field around a wire. It can be made even stronger by coiling up the wire many times. A coil of wire that is used to produce a magnetic field is called an **electromagnet**.

Use a battery for this experiment. *Never* connect anything to mains electricity.

Use plastic covered wire.

Use a wire, a battery and an iron nail to make an electromagnet. Wind the wire around the nail many times. Connect each end of the wire to the battery. Now use the nail to pick up pins. See how they fall off if you disconnect the battery.

Trains that float

Electromagnets are used on some special trains instead of wheels. The magnetic force from electromagnets hold the train a few centimetres above the track, and push it along.

These trains do not touch the track, so there is no friction. They can go very fast.

Electric motors

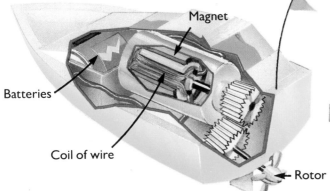

Magnet

Batteries

Coil of wire

Rotor

An **electric motor** works using electromagnetism. It has a coiled up wire inside that sits between the poles of a magnet. When current flows through the coil, a magnetic field is produced. This makes the rotor spin around.

Generating electricity

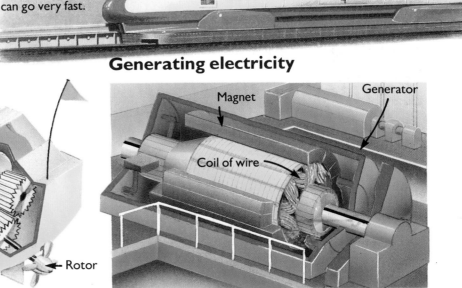

Generator

Magnet

Coil of wire

An electric current is produced inside a wire, if it is moved in a magnetic field. This is how a **generator**, or **dynamo**, produces electricity. An engine spins a coil of wire between the poles of a magnet. This generates electric current.

Records and tapes

What makes a record player work? How is sound recorded on to a tape? How does a loudspeaker work? How does a microphone pick up sound?

Electromagnetism makes all these things work. It is used to record sound on to records and tapes and to play back sound through loudspeakers.

Microphones

Microphones change sounds into electrical signals. A thin disc inside a microphone vibrates as sound waves* hit it. This makes a coil of wire vibrate.

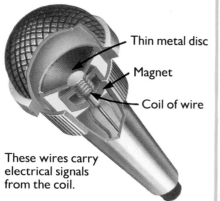

Thin metal disc

Magnet

Coil of wire

These wires carry electrical signals from the coil.

The coil sits between the poles of a magnet*. As the coil moves, it produces an electric current. The current flows backwards and forwards in time with the sound and is carried down a wire to an amplifier*.

How does a loudspeaker work?

A loudspeaker works in the opposite way to a microphone. It changes electrical signals back into sound waves.

The electrical signals from a hi-fi make a thin plastic or paper cone vibrate. This produces the sounds you hear.

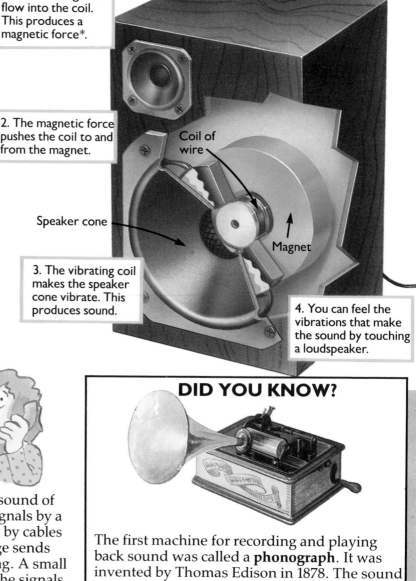

1. Electrical signals flow into the coil. This produces a magnetic force*.

2. The magnetic force pushes the coil to and from the magnet.

Coil of wire

Speaker cone

Magnet

3. The vibrating coil makes the speaker cone vibrate. This produces sound.

4. You can feel the vibrations that make the sound by touching a loudspeaker.

How telephones work

When you make a telephone call, the sound of your voice is changed into electrical signals by a microphone. These signals are carried by cables to a **telephone exchange**. The exchange sends the signals to the person you are calling. A small loudspeaker in their telephone turns the signals back into sound waves.

DID YOU KNOW?

The first machine for recording and playing back sound was called a **phonograph**. It was invented by Thomas Edison in 1878. The sound was recorded on to a drum covered in tin foil.

Playing records

A record has a thin groove that runs from the outside to the centre. There are millions of tiny bumps in it. When you play a record, a tiny crystal, called a **stylus**, runs along the groove.

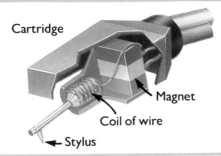

Cartridge

Magnet

Coil of wire

Stylus

The stylus moves up and down over the bumps. This makes a tiny coil vibrate inside the **cartridge**. The coil sits between the poles of a magnet, so it produces electrical signals as it vibrates. The signals go along wires to an amplifier.

Amplifiers

The electrical signals from record players and cassette decks are too weak to make loudspeakers work on their own. An **amplifier** is used to make the signals stronger. It is connected to the loudspeakers. When you turn the volume control up, the amplifier makes the signals stronger.

Record deck

Amplifier

Cassette deck

Cassette recorders

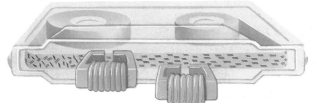

The tape inside a cassette is covered by lots of tiny magnets. To record sound on tape, electrical signals are sent to an electromagnet, called the **recording head**. This arranges the magnets in a special pattern that matches the music.

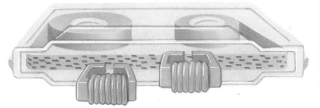

When you play a cassette, the tape moves past the **playback head**. This picks up the pattern on the tape and changes it into electrical signals. The electrical signals go through an amplifier to loudspeakers that change them back into sound.

Video recorders

A video tape recorder works in the same way as a normal cassette recorder. A strip along one edge of the tape is used to record sound, while the picture is recorded along the middle of the tape.

Producing electricity

Many things work using electricity from wall sockets. A light goes on as soon as you press a light switch. Where does the electricity come from and how does it reach your home?

Electricity is produced in power stations. Most of them produce electricity by burning coal, oil or gas. Others use the energy from nuclear power*, moving water or wind*.

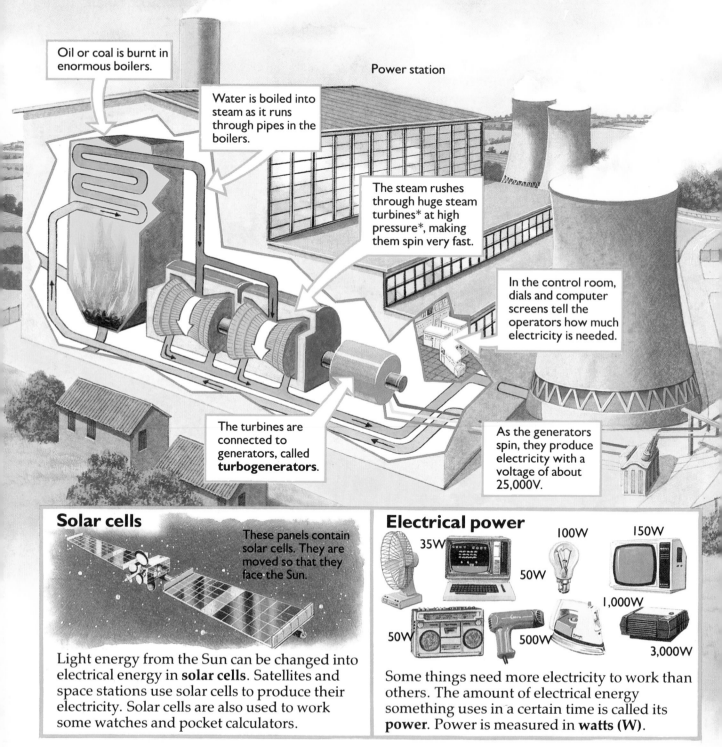

Oil or coal is burnt in enormous boilers.

Water is boiled into steam as it runs through pipes in the boilers.

Power station

The steam rushes through huge steam turbines* at high pressure*, making them spin very fast.

In the control room, dials and computer screens tell the operators how much electricity is needed.

The turbines are connected to generators, called **turbogenerators**.

As the generators spin, they produce electricity with a voltage of about 25,000V.

Solar cells

These panels contain solar cells. They are moved so that they face the Sun.

Light energy from the Sun can be changed into electrical energy in **solar cells**. Satellites and space stations use solar cells to produce their electricity. Solar cells are also used to work some watches and pocket calculators.

Electrical power

35W   100W   150W   50W   50W   500W   1,000W   3,000W

Some things need more electricity to work than others. The amount of electrical energy something uses in a certain time is called its **power**. Power is measured in **watts (W)**.

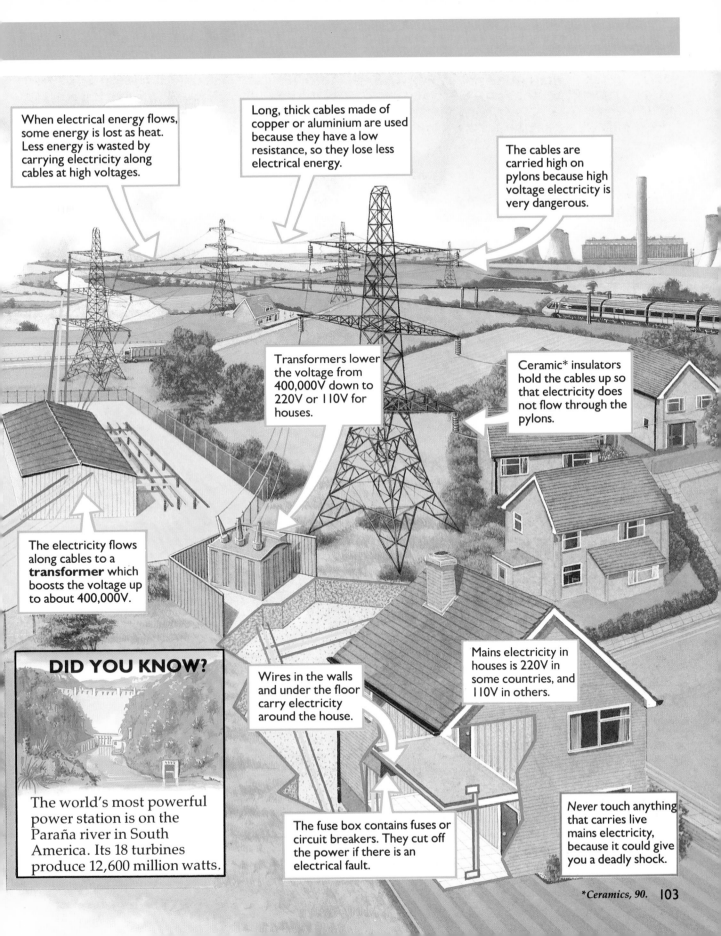

When electrical energy flows, some energy is lost as heat. Less energy is wasted by carrying electricity along cables at high voltages.

Long, thick cables made of copper or aluminium are used because they have a low resistance, so they lose less electrical energy.

The cables are carried high on pylons because high voltage electricity is very dangerous.

Transformers lower the voltage from 400,000V down to 220V or 110V for houses.

Ceramic* insulators hold the cables up so that electricity does not flow through the pylons.

The electricity flows along cables to a **transformer** which boosts the voltage up to about 400,000V.

DID YOU KNOW?

The world's most powerful power station is on the Paraña river in South America. Its 18 turbines produce 12,600 million watts.

Wires in the walls and under the floor carry electricity around the house.

Mains electricity in houses is 220V in some countries, and 110V in others.

The fuse box contains fuses or circuit breakers. They cut off the power if there is an electrical fault.

Never touch anything that carries live mains electricity, because it could give you a deadly shock.

The electromagnetic spectrum

Light is made up of waves called electromagnetic waves. Apart from light, there are many other kinds of electromagnetic waves, but they are all invisible. Together they form the electromagnetic spectrum.

All electromagnetic waves travel at 300,000km per second and can travel through a vacuum. Electromagnetic waves with different wavelengths and frequencies can be used for different things.

Gamma rays

Gamma rays come from nuclear radiation*. They can travel through many things, even metal. Gamma rays are very dangerous because they kill living cells, but they are used in small amounts to help cure some diseases.

Ultraviolet waves

Ultraviolet (UV) radiation from the Sun gives people a suntan by making their skin produce a brown chemical, called **melanin**.

Too much UV radiation is bad for you. A gas called **ozone** in the atmosphere cuts out some of the Sun's UV radiation. People are worried because pollution is destroying this gas.

Short wavelength
High frequency

X-rays

X-rays are used to look inside people. They can only travel through soft things, so hard things, like bone, show up as shadows. X-rays are also used in airports to check what may be hidden in people's suitcases.

Visible light

The light that you see, called **visible light**, is only a small part of the electromagnetic spectrum. Visible light with different wavelengths produces different colors*.

DID YOU KNOW?

Electromagnetic waves are made of changing electric and magnetic fields*. The first person to understand the link between electricity and magnetism was James Clerk Maxwell, in 1864.

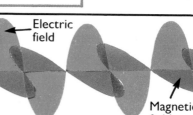

Electric field

Magnetic field

*Colored light, 61; Electric and magnetic fields, 99; Nuclear radiation, 77.

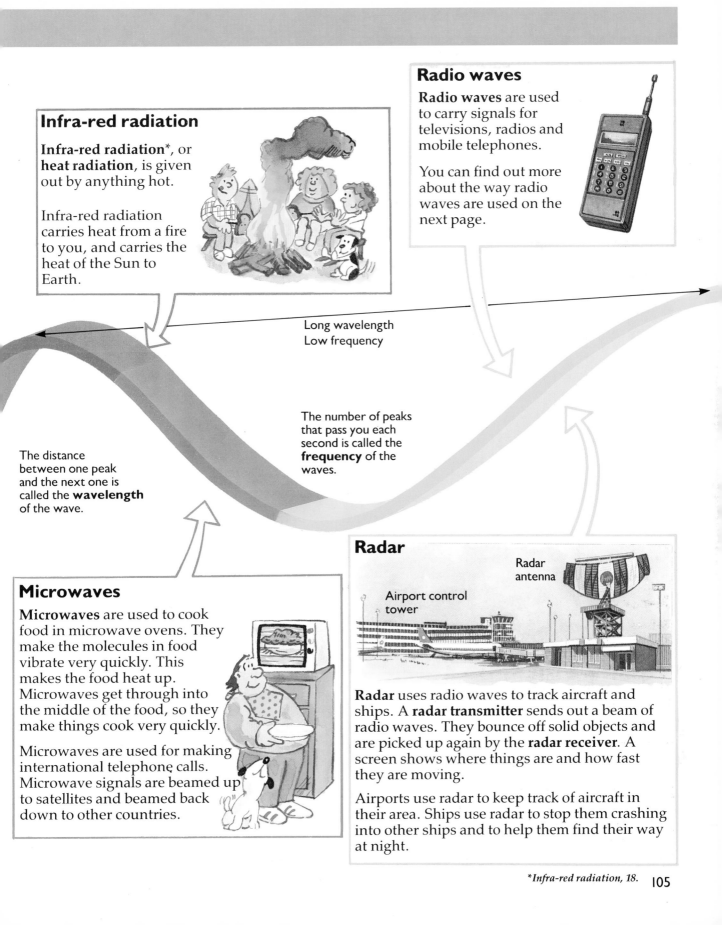

Infra-red radiation

Infra-red radiation*, or **heat radiation**, is given out by anything hot.

Infra-red radiation carries heat from a fire to you, and carries the heat of the Sun to Earth.

Radio waves

Radio waves are used to carry signals for televisions, radios and mobile telephones.

You can find out more about the way radio waves are used on the next page.

Long wavelength
Low frequency

The number of peaks that pass you each second is called the **frequency** of the waves.

The distance between one peak and the next one is called the **wavelength** of the wave.

Microwaves

Microwaves are used to cook food in microwave ovens. They make the molecules in food vibrate very quickly. This makes the food heat up. Microwaves get through into the middle of the food, so they make things cook very quickly.

Microwaves are used for making international telephone calls. Microwave signals are beamed up to satellites and beamed back down to other countries.

Radar

Radar antenna

Airport control tower

Radar uses radio waves to track aircraft and ships. A **radar transmitter** sends out a beam of radio waves. They bounce off solid objects and are picked up again by the **radar receiver**. A screen shows where things are and how fast they are moving.

Airports use radar to keep track of aircraft in their area. Ships use radar to stop them crashing into other ships and to help them find their way at night.

Radio and television

There are radio waves all around you, but you cannot hear or see them. They are picked up by radios that turn them into sound waves and by televisions that turn them into sound waves and light waves.

The sounds you hear on a radio may have travelled thousands of kilometres to reach you. Radio waves travel at the speed of light. This is why people living far away from each other can hear the same radio show at the same time.

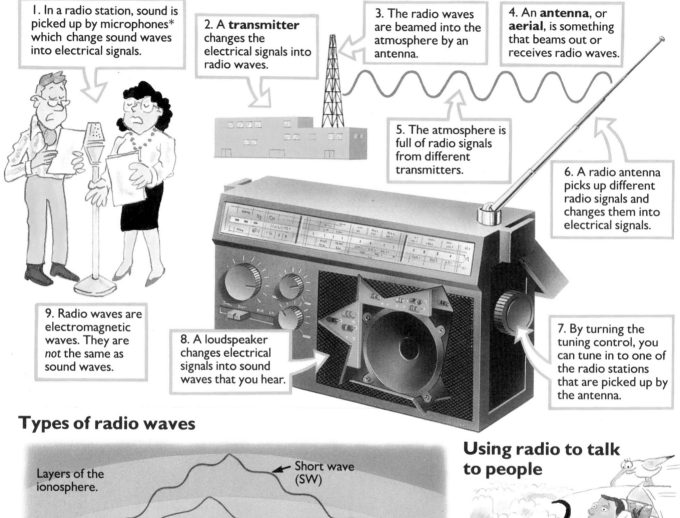

1. In a radio station, sound is picked up by microphones* which change sound waves into electrical signals.

2. A **transmitter** changes the electrical signals into radio waves.

3. The radio waves are beamed into the atmosphere by an antenna.

4. An **antenna**, or **aerial**, is something that beams out or receives radio waves.

5. The atmosphere is full of radio signals from different transmitters.

6. A radio antenna picks up different radio signals and changes them into electrical signals.

7. By turning the tuning control, you can tune in to one of the radio stations that are picked up by the antenna.

8. A loudspeaker changes electrical signals into sound waves that you hear.

9. Radio waves are electromagnetic waves. They are *not* the same as sound waves.

Types of radio waves

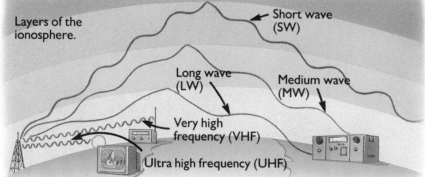

Layers of the ionosphere.

Short wave (SW)

Long wave (LW)

Medium wave (MW)

Very high frequency (VHF)

Ultra high frequency (UHF)

There are different types of radio waves. Long, medium and short waves can travel very far. This is because they bounce off layers in the atmosphere, called the **ionosphere**. VHF radio waves and the UHF waves that carry television signals travel short distances because they cannot bounce off the ionosphere.

Using radio to talk to people

Two-way radios are able to send out and receive radio signals. They are used by many people such as taxi drivers, the police and aircraft pilots.

Some telephones, called **mobile telephones**, use radio waves instead of telephone wires.

Radio waves carry sound and pictures to televisions in people's homes. A television set turns the radio waves into light waves and sound waves that you see and hear.

How does television work?

Television cameras pick up light from things in the studio. They divide the light into primary colors* and then change it into electrical signals. The signals are changed into radio waves and sent out by a transmitter.

1. The antenna picks up the radio waves and changes them into electrical signals.

2. The main part of a television set is called a **cathode ray tube**. This is the screen you watch.

5. There are chemicals, called **phosphors**, on the inside of the screen. They glow when the electrons hit them.

6. Three kinds of phosphor glow red, green and blue. All the different colors in the television picture are made by mixing these three colors.

3. The picture is built up by beams of electrons that scan across the screen, moving down line by line. This happens so quickly that you cannot see the moving beam.

4. There are three electron beams, one for each primary color, red, green and blue.

Cable television

Some television channels are transmitted by electrical signals that flow along a special cable to people's homes. This is called **cable television**.

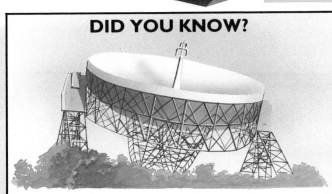

DID YOU KNOW?

The Sun and other stars send out radio waves through Space. They are picked up by huge dish antennas, called **radio telescopes**. Astronomers use them to find out about distant galaxies.

Satellite television

You can watch television programs from all over the world if you have a satellite receiver. Programs are sent out, using microwaves*, to an orbiting satellite. The satellite then beams them down to people's houses in other countries.

*Microwaves, 105; Primary colors, 62.

Computer technology

Computers can do many different things. They are used to send rockets into Space, to forecast the weather, to make robots work, for typing letters, playing games and making music.

Computers can store huge amounts of information that would fill thousands of pages if it was all written on paper. In only a few seconds, they can find any of the information stored in them.

Computers

In just one second, computers can do millions of calculations that would take people weeks or even years to do. But computers cannot think for themselves.

Computers need to be told what to do. They are given a list of instructions, called a **computer program**. Computer programs are written in special languages, such as **BASIC** or **LOGO**.

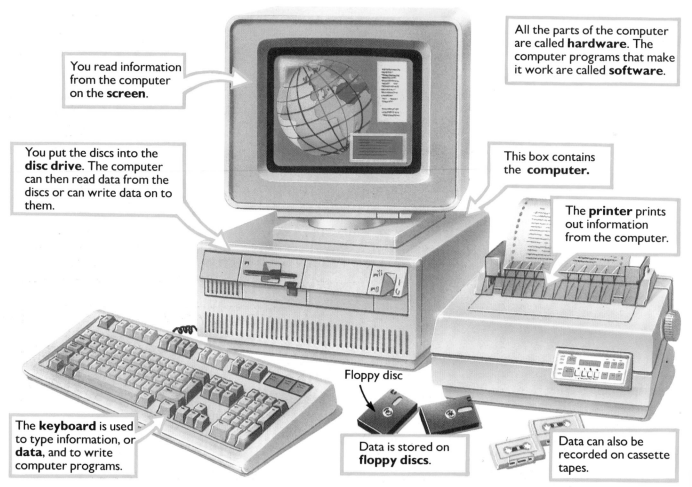

You read information from the computer on the **screen**.

All the parts of the computer are called **hardware**. The computer programs that make it work are called **software**.

You put the discs into the **disc drive**. The computer can then read data from the discs or can write data on to them.

This box contains the **computer.**

The **printer** prints out information from the computer.

The **keyboard** is used to type information, or **data**, and to write computer programs.

Floppy disc

Data is stored on **floppy discs**.

Data can also be recorded on cassette tapes.

How computers work?

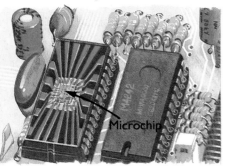

Microchip

Computers work using **microchips**, which are also called **silicon chips**. The microchips are the 'brains' of the computer. They contain lots of tiny **electronic circuits** that are able to store information and do calculations.

Digital information

Computers store all information in the form of numbers. They use numbers made up of just ones and zeros to make codes that stand for letters, numbers, sounds and pictures. Information that is stored in this way is called **digital information**.

Computers only recognize numbers made of ones and zeros, called binary numbers*. This is because their microchips work using lots of tiny switches. The number 'one' stands for 'switched on' and 'zero' stands for 'switched off'.

Lasers

Lasers produce a fine beam of light that does not spread out like ordinary light. It is the brightest light known, even brighter than sunlight. Laser light has so much energy that it can even cut through metal.

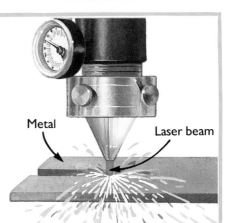

Metal

Laser beam

Laser printer

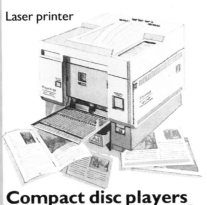

Lasers are used in many ways. They are used to carry computer messages and telephone calls through optic fibers*. They are also used to test people's eyesight, to print newspapers, to measure distances accurately, and by surgeons in operations.

Compact disc players

Compact disc

Sound is recorded on a **compact disc** by changing it into digital information, like computer data. A compact disc player contains a laser. The laser beam reads the digital information from the disc so that it can then be changed back into sound waves.

DID YOU KNOW?

Holograms are special three-dimensional photographs taken with laser beams. The pictures they show look like real objects because you see different views as you walk around them.

Scientists and inventors

Ampère, André 1775-1836
French physicist who first realized that electricity and magnetism are linked. The unit of electric current, the ampere, is named after him.

Archimedes c. 287-212 BC
Greek scientist who first understood how pressure changes with depth in liquids and gases. He developed the theory of pulleys and levers, but is best known for Archimedes' principle, which explains how things float.

Aristotle 384-322 BC
Greek philospher who founded modern scientific thinking. He thought that all things were made of fire, earth, air and water. He believed that the Earth was at the center of the Universe and that the Universe was a sphere. These ideas were later disproved.

Babbage, Charles 1792-1871
English mathematician who built a mechanical calculating machine, called the analytical engine. His ideas formed the basis for electronic computers.

Baird, John Logie 1888-1946
Scottish inventor who, in 1926, first demonstrated television. He opened the first television studio in 1929.

Becquerel, Antoine 1852-1908
French physicist who discovered nuclear radioactivity in 1896.

Bell, Alexander Graham 1847-1922
Scottish inventor of the telephone. He invented many things, including deaf aids.

Benz, Karl 1844-1929
German engineer who invented the first practical motor car that was powered by an internal combustion engine.

Bohr, Niels 1885-1962
Danish physicist who, in 1913, introduced a new theory which changed people's understanding of the structure of atoms.

Boyle, Robert 1627-1691
Anglo-Irish philosopher who was the first person to suggest that things were made of simple elements. This contradicted Aristotle's ideas. He also made major discoveries about gases.

Braun, Wernher von 1912-1977
German engineer who developed the first long-range rocket, called the V2 missile.

Carothers, Wallace 1896-1937
American chemist who discovered nylon, the first man-made polymer fiber to be widely used.

Cayley, George 1773-1857
English inventor whose ideas led to the invention of the airplane. He built the first glider that carried a person.

Chadwick, James 1891-1974
English physicist who discovered the neutron inside atoms.

Cierva, Juan de la 1895-1936
Spanish engineer who invented the autogyro, one of the earliest types of helicopter.

Copernicus, Nicolaus 1473-1543
Polish astronomer who correctly thought that the Earth orbited the Sun. Until then, people had believed that the Sun orbited the Earth.

Curie, Marie 1867-1934 and **Pierre** 1859-1906
French scientists who discovered the radioactive elements, radium and polonium.

Daguerre, Louis 1787-1851
French painter and designer who invented the first practical photographic process.

Daimler, Gottlieb 1834-1900
German engineer who developed the first successful internal combustion engine. It worked using gasoline fuel.

Dalton, John 1766-1844
English chemist who established the theory that everything is made of atoms.

Diesel, Rudolf 1858-1913
German engineer who developed a type of internal combustion engine called the diesel engine.

Dunlop, John 1840-1921
Scottish veterinary surgeon who invented the first air-filled, or pneumatic, tire.

Eastman, George 1854-1932
American industrialist who, in 1888, invented the first flexible roll film for use in the first Kodak camera. Until then, photographs were taken on individual glass plates.

Edison, Thomas 1847-1931
American scientist who produced more than 1,000 inventions. He invented the light bulb and the phonograph, which was the first record player.

Einstein, Albert 1879-1955
German physicist who worked out the theories of relativity. They explain what happens when things travel close to the speed of light. He also showed that mass could be changed into energy, which led to the discovery of nuclear energy.

Faraday, Michael 1791-1867
English scientist who invented the electric motor, the dynamo and the transformer. He was the first person to discover many compounds containing carbon and chlorine.

Fermat, Pierre de 1601-1665
French mathematician who founded the modern theory of numbers.

Fermi, Enrico 1901-1954
Italian physicist who designed and built the first nuclear reactor.

Fleming, Alexander 1881-1955
Scottish scientist who discovered penicillin.

Fox Talbot, William 1800-1877
English scientist who invented photographic negatives from which many prints could be made.

Franklin, Benjamin 1706-1790
American scientist and politician. He invented lightning conductors which protect buildings by carrying lightning safely down to the ground.

Gabor, Dennis 1900-1979
Hungarian physicist who invented holograms.

Galilei, Galileo 1564-1642
Italian scientist. He discovered the way pendulums work and showed how gravity affects falling things. He was one of the first people to look at the solar system using a telescope and discovered many things, such as Jupiter's moons. He also invented the thermometer.

Goddard, Robert 1882-1945
American physicist who was one of the pioneers of space rocket design. He launched the first liquid-fuelled rocket in 1926.

Gutenberg, Johannes 1400-1468
German printer who introduced the first printing press into Europe.

Hero of Alexandria
1st century AD
Greek-Egyptian engineer and mathematician who explained how siphons, pumps and fountains work. He also invented the first steam-powered machine, a spinning metal sphere.

Hertz, Heinrich 1857-1894
German physicist who discovered electromagnetic waves and founded the principles of radio transmission. In the late 1880s, he was the first person to show that electromagnetic waves travel at the speed of light. He also showed that they can be reflected and refracted.

Huygens, Christiaan 1629-1695
Dutch physicist, astronomer and mathematician. He built the first pendulum clock, improved the telescope, discovered Saturn's rings, and was the first person to suggest that light was made up of waves.

Joule, James 1818-1889
English scientist who studied heat and energy. He developed the law of conservation of energy with W. Thomson. This law says that you can never get more energy out than you put in. The unit of energy, the joule, is named after him.

Kepler, Johannes 1571-1630
German astronomer who worked out the way that planets move in the solar system. He was the first person to suggest that they moved in oval orbits rather than circles.

Lavoisier, Antoine 1743-1794
French chemist who discovered the role of oxygen in breathing and burning. He also introduced one of the first systems for naming chemicals.

Leclanché, Georges 1839-1882
French inventor of the first kind of dry-cell battery. Dry-cell batteries are used in radios and flashlights.

Lenoir, Étienne 1822-1900
Belgian engineer who invented the first gas-powered internal combustion engine.

Lilienthal, Otto 1848-1896
German engineer who designed and built gliders. He was one of the pioneers of the basic ideas of manned flight.

Lippershey, Hans c.1570-c.1619
Dutch spectacle maker who invented the telescope.

Lodge, Oliver 1851-1940
English physicist who, at about the same time as Marconi, showed that radio waves could be used for signalling.

Lumière, Auguste 1862-1954
and **Louis** 1864-1948
French inventors who developed the movie camera and color photography. They were the first people to open a public movie. The first film was shown in it in 1895.

Mach, Ernst 1838-1916
Czechoslovakian physicist who worked out the 'Mach numbers', the speed of an object compared to the speed of sound in air.

Marconi, Guglielmo 1874-1937
Italian inventor who developed the first radio transmitters and receivers. In 1901, he transmitted the first radio signals across the Atlantic.

Maxwell, James Clerk
1831-1879
Scottish scientist whose theory of electromagnetic radiation predicted the existence of radio waves. He was the first person to realize that light is a type of electromagnetic radiation.

Mendel, Gregor 1822-1884
Austrian monk who founded the science of genetics which explains how the characteristics of parents are handed down to their children.

Mendeleyev, Dmitry 1834-1907
Russian chemist who worked out the Periodic Table of the elements which forms the basis of modern chemistry.

Montgolfier, Joseph 1740-1810
and **Jacques** 1745-1799
French brothers who invented the hot-air balloon. Their balloon carried the first person into the air, in 1783.

Morse, Samuel 1791-1872
American inventor who developed the electric telegraph in the USA and invented Morse code.

Newcomen, Thomas 1663-1729
English inventor of the first practical steam engine, which first began working in 1712.

Newton, Isaac 1642-1727
English scientist who worked out the laws of motion, the theory of gravitation and many new mathematical theories. He discovered that white light is made up of all the colors of the spectrum and invented the reflector telescope. He is recognized as one of the most original thinkers of all time. The unit of force, the newton, is named after him.

Nipkow, Paul 1860-1940
German inventor who was one of the pioneers of television.

Nobel, Alfred 1833-1896
Swedish chemist who invented dynamite. He founded the Nobel Prizes which are given to people who make advances in physics, chemistry, medicine, literature and world peace.

Oersted, Hans 1777-1851
Danish scientist who discovered that an electric current produces magnetism. He was one of the first people to understand electromagnetism.

Otto, Nikolaus 1832-1891
German engineer who built the first four-stroke internal combustion engine.

Pascal, Blaise 1623-1662
French mathematician who invented the mechanical adding machine. He worked out many mathematical theories, including the theory of probability.

Planck, Max 1858-1947
German physicist who worked out quantum theory, which changed people's understanding of energy and led to many new discoveries.

Priestley, Joseph 1733-1804
English scientist who discovered oxygen in 1774. He was also the inventor of the first fizzy drink.

Ptolemy 2nd century AD
Greek scientist and astronomer. He believed that the Sun and planets orbited the Earth in a series of complicated circles. This was only disproved in the 16th century.

Röntgen, Wilhelm 1845-1923
German physicist who discovered X-rays.

Rutherford, Ernest 1871-1937
New Zealand scientist who first suggested that atoms have a nucleus in the center around which the electrons move.

Sikorsky, Igor 1889-1972
Russian-American engineer who designed the first modern helicopter.

Stephenson, George 1781-1848
English engineer who developed steam locomotives.

Swan, Joseph 1828-1914
English scientist who invented the light bulb at about the same time as Edison.

Tesla, Nikola 1856-1943
Yugoslavian-American scientist. He invented a type of electric motor called an induction motor.

Thomson, Joseph 1856-1940
English physicist who discovered the electron.

Thomson, William (Lord Kelvin) 1824-1907
Irish physicist who developed the science of thermodynamics, which studies the link between heat and other forms of energy.

Torricelli, Evangelista 1608-1647
Italian scientist who invented the barometer.

Vinci, Leonardo da 1452-1519
Italian artist and inventor. Many of the devices he invented were so far ahead of his time that they were not developed till hundreds of years later.

Volta, Alessandro 1745-1827
Italian physicist who invented the first battery.

Watson-Watt, Robert 1892-1973
Scottish scientist who developed radar.

Watt, James 1736-1819
Scottish engineer who developed and improved Newcomen's steam engine. The unit of power, the watt, is named after him.

Whittle, Frank 1907-
English engineer who invented the jet engine.

Wright, Wilbur 1867-1912 and **Orville** 1871-1948
American brothers who built the first successful powered airplane. It first flew at Kitty Hawk, USA, in 1903.

Zeppelin, Ferdinand von 1838-1917
German inventor who built the first airship.

Zworykin, Vladimir 1889-1982
Russian-American engineer who was one of the pioneers of television.

Charts and tables

The solar system

Planet	Distance from Sun	Diameter	Time taken to orbit Sun	Number of moons
Mercury	58 million km	4,900km	88 days	0
Venus	108 million km	12,100km	225 days	0
Earth	150 million km	12,756km	365. 25 days	1
Mars	228 million km	6,800km	687 days	2
Jupiter	778 million km	143,000km	11.86 years	14
Saturn	1,431 million km	120,000km	29.46 years	17
Uranus	2,886 million km	51,000km	84 years	9
Neptune	4,529 million km	49,000km	165 years	2
Pluto	5,936 million km	3,000km	248 years	1

Earth facts

Diameter at Equator	12,756km
Diameter at Poles	12,712km
Greatest height above sea level (Mt. Everest)	8,848m
Greatest depth below sea level (Marianas Trench)	11,033m
Land area	149 million km^2
Ocean area	361 million km^2
Amount of Earth covered by sea	71%

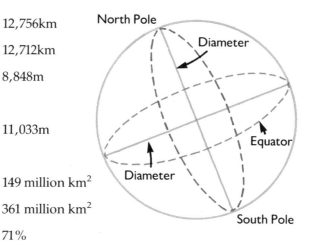

North Pole

Diameter

Equator

Diameter

South Pole

Sun facts

Diameter	1,400,000km
Temperature at center	16 million °C
Temperature at surface	5,500 °C
Time taken for sunlight to reach Earth	8 minutes, 20 seconds

Temperature chart

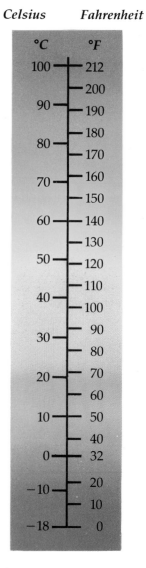

Celsius   *Fahrenheit*

°C	°F
100	212
	200
90	190
	180
80	170
70	160
	150
60	140
	130
50	120
	110
40	100
30	90
	80
20	70
	60
10	50
	40
0	32
−10	20
	10
−18	0

Celsius and Fahrenheit are units for measuring temperature. To change Celsius into Fahrenheit, you multiply the temperature in Celsius by 9, divide the answer by 5, and add 32.

To change Fahrenheit into Celsius, subtract 32 from the temperature in Fahrenheit, multiply by 5, and then divide the answer by 9.

Metric units

Length

1 centimetre (cm) = 10 millimetres (mm)
1 metre (m) = 100 centimetres
1 kilometre (km) = 1,000 metres

Mass

1 kilogram (kg) = 1,000 grams (g)
1 tonne (t) = 1,000 kilograms

Area

100 square millimetres (mm^2) = 1 square centimetre (cm^2)
1 square metre (m^2) = 10,000 square centimetres
1 hectare = 10,000 square metres
1 square kilometre (km^2) = 1 million square metres

Volume

1 cubic centimetre (cc) = 1 millilitre (ml)
1 litre (l) = 1,000 millilitres
1 cubic metre (m^3) = 1,000 litres

Imperial units

Length

1 foot (ft) = 12 inches (in)
1 yard (yd) = 3 feet
1 mile = 1,760 yards

Mass

1 pound (lb) = 16 ounces (oz)
1 ton = 2,240 pounds

Area

1 square foot (ft^2) = 144 square inches (in^2)
1 square yard (yd^2) = 9 square feet
1 acre = 4,840 square yards
1 square mile = 640 acres

Volume

1 gallon = 8 pints
1 cubic foot (ft^3) = 7.48 gallons
1 gallon (US) = 0.83 gallons (UK)

Metric prefixes

The prefixes written before a unit of measurement tells you how much the unit is multiplied by.

For example, 1 kilovolt (1kV) equals one thousand volts (1,000V).

Prefix	micro	milli	centi	deci	kilo	mega
Symbol	μ	m	c	d	k	M
Means unit is multiplied by	one millionth (0.000,001)	one thousandth (0.001)	one hundredth (0.01)	one tenth (0.1)	one thousand (1,000)	one million (1,000,000)

Metric units into imperial units

To convert	into	multiply by
Length		
Centimetres	inches	0.39
Metres	feet	3.28
Kilometres	miles	0.62
Area		
Square metres	square feet	10.76
Hectares	acres	2.47
Square kilometres	square miles	0.39
Volume		
Cubic metres	cubic feet	35.32
Litres	pints	1.76
Litres	gallons	0.22
Mass		
Grams	ounces	0.04
Kilograms	pounds	2.21
Tonnes	tons	0.98

Imperial units into metric units

To convert	into	multiply by
Length		
Inches	centimetres	2.54
Feet	metres	0.31
Miles	kilometres	1.61
Area		
Square feet	square metres	0.09
Acres	hectares	0.41
Square miles	square kilometres	2.59
Volume		
Cubic feet	cubic metres	0.03
Pints	litres	0.57
Gallons	litres	4.55
Mass		
Ounces	grams	28.35
Pounds	kilograms	0.45
Tons	tonnes	1.02

Glossary

Absorb. To take in. For example, a sponge takes in, or absorbs, water.

Acceleration. The change of velocity caused by changing speed or direction.

Acoustics. The way sound travels in a room. It also means 'the science of sound'.

Aerodynamics. The study of the way air flows around things.

Aerofoil. The special shape of a wing that enables aircraft to lift off the ground.

Aerosol. A pressurized container used to spray liquids, like paint, in a fine mist.

Air resistance. The push of air against moving things that works to slow them down.

Alchemy. An ancient form of chemistry. Alchemists tried to turn things into gold.

Alloy. A metal that is made of a mixture of different metals.

Alternating current. An electric current that changes direction, usually many times each second.

Amplifier. An electronic device that boosts electrical signals.

Antenna. A length of wire that is used to send out or receive radio waves. Also called an aerial.

Astronomy. The scientific study of the objects in Space, such as stars, asteroids and planets.

Atmosphere. The blanket of gases that surrounds the Earth.

Atmospheric pressure. The pressure on the Earth caused by the weight of the gases in the atmosphere.

Atoms. Tiny particles out of which everything is made.

Attract. To make something come closer. A magnet attracts iron.

Barometer. A device for measuring atmospheric pressure.

Biology. The scientific study of all living things.

Boiling point. The temperature at which a liquid boils, changing into a gas.

Botany. The scientific study of plants.

Buoyancy. The ability of something to float.

Calorie. A unit of energy that is often used to measure the energy content of food. One calorie equals 4.18 joules.

Celsius. The temperature scale which sets the melting point of pure water at zero degrees and its boiling point at 100 degrees.

Centigrade scale. A scale that is divided into 100 units.

Chemical reaction. When atoms of different things join up together to make something new.

Chemistry. The scientific study of all substances, and how they react and combine together.

Chlorophyll. The chemical inside leaves that gives green plants their color and is essential for photosynthesis.

Circuit. A continuous electrical conductor, such as a wire, through which electricity flows.

Combustion. The process of burning.

Compact disc. A disc on which sound or computer data is recorded in digital form.

Compass. A device that uses the Earth's magnetism to show you which direction you are facing.

Compound. A substance made of atoms of different elements chemically joined together.

Compress. To squash up into a smaller space.

Condensation. The way a vapor changes into a liquid as it cools down.

Conductor. A material that allows either electricity or heat to flow through it easily.

Contraction. Shrinking to a smaller size.

Convection. The way heat is carried by a liquid or a gas. A convection current is a flow of liquid or gas carrying heat.

Crude oil. Oil that is extracted from the inside of the Earth, before it has been refined.

Current electricity. The type of electricity that can flow through wires.

Data. Another word for information.

Decelerate. To slow down.

Decibel. A unit for measuring the loudness, or intensity, of sound.

Density. How much something weighs for its size.

Diameter. The straight line between opposite sides of a circle that passes through its centre.

Digital information. Information that is stored in computers as binary numbers.

Diffusion. The way that molecules of one material spread through those of another.

Direct current. Electric current that flows in only one direction around a circuit.

Dynamo. A type of generator that produces direct current.

Earthquake. Vibrations of the Earth's crust caused by the movement of rock beneath its surface.

Echo-location. Navigation using echoes from high-pitched sounds.

Efficiency. The amount of energy you get out compared to how much energy you put in.

Electric charge. Something that has an electric charge carries electricity. There are two types of electric charge, called positive and negative.

Electric field. The space around an electric charge in which its electric force acts.

Electromagnet. A coil of wire that produces a magnetic field when an electric current is passed through it.

Electrons. Tiny particles inside atoms that carry a negative charge.

Electronics. Technology to do with circuits and microchips.

Element. A substance that is made of only one type of atom.

Energy. The ability to do work and provide power. There are many different types of energy including heat energy, light energy, sound energy, chemical energy and nuclear energy.

Evaporation. The way a liquid changes into a gas when it is below its boiling point.

Fahrenheit. The temperature scale which sets the freezing point of pure water at 32 degrees and its boiling point at 212 degrees.

Fission. The splitting up of the nucleus of an atom, producing a huge amount of nuclear energy.

Fluid. Either a liquid or a gas.

Focus. The point where rays of light from a lens or a curved mirror meet.

Force. A push or pull that makes things move, change shape or change direction.

Fossil fuel. Fuel, like coal or oil, that has formed over millions of years and is dug up from the ground.

Freezing point. The temperature at which a liquid freezes, changing into a solid.

Friction. The force that works to stop things moving, or slow them down if they are already moving.

Fusion. The joining up of the nuclei of different atoms, producing a huge amount of nuclear energy.

Generator. A machine that turns the energy of movement, or kinetic energy, into electrical energy.

Geography. The scientific study of the Earth.

Geology. The scientific study of the Earth's rocks and crust.

Graphite. Soft, flaky form of the element carbon, used as a lubricant and in pencil leads.

Heat radiation. The way that heat is carried by infra-red rays.

Hologram. A special three-dimensional photograph that is taken with a laser.

Humidity. The amount of moisture there is in the air.

Hydroelectricity. Electricity that is generated from the energy of moving water.

Ignite. To set on fire.

Inertia. The tendency of things to keep still, or to carry on moving at the same speed in the same straight line, unless they are acted on by a force.

Inflammable. Easily set on fire.

Insulator. A material that does not allow heat or electricity to pass through it easily.

Kerosene. The fuel that is used in jet engines. Also called paraffin.

Laser. A device which produces a very bright beam of light. It can be used for cutting things or carrying information.

Lodestone. Naturally magnetic material. Also called magnetite.

Luminous. A luminous object is something that gives out light.

Lubrication. The use of a thick liquid, called a lubricant, to reduce friction between the moving parts of a machine.

Magnet. A material which produces a magnetic force that attracts the metals iron, cobalt or nickel. If a magnet is held freely, it lines up with the North and South Poles of the Earth.

Magnetic field. The space around a magnet in which its magnetic force works.

Magnify. To make bigger using lenses.

Mass. The amount there is of something.

Mathematics. The science of numbers, quantities and shapes.

Matter. Any physical thing that takes up space.

Melting point. The temperature at which a solid melts, turning into a liquid.

Meteorology. The scientific study of the weather.

Microchip. A tiny piece of silicon that contains thousands of electronic circuits. They are also called silicon chips.

Microscope. A device that uses lenses to magnify small things many times.

Mixture. Two or more elements or compounds that are mixed up together, but are not chemically joined up to each other.

Molecule. A particle that contains two or more atoms joined together.

Neutrons. Particles inside the nuclei of atoms. They carry no electric charge.

Nuclear force. A very strong force which holds protons and neutrons together inside the nucleus of an atom.

Nuclear radiation. Dangerous radiation that is given out by radioactive materials.

Nuclear reactor. The place where the nuclei of atoms are split in order to release energy.

Nucleus. The central part of an atom which contains protons and neutrons. The electrons whiz around the nucleus.

Opaque. An opaque object does not let any light pass through it.

Orbit. The path of a satellite or a planet, often circular or oval in shape.

Optic fiber. A thin piece of glass that can carry light long distances. The light can be used to carry telephone calls and computer data.

Ozone. A layer of gas in the atmosphere that protects the Earth from the Sun's ultraviolet radiation.

Particle. A tiny piece of a material.

Pendulum. A hanging weight that swings backwards and forwards. It is used as a regulator for clocks because each swing lasts the same time.

Periodic Table. A table that shows all the elements arranged in groups. The elements in each group have similar properties.

Perspective. A way of drawing pictures to give a feeling of distance and depth.

Physics. The scientific study of matter and energy.

Pitch. How high or low a sound is.

Planet. A large ball of rock or gas which orbits a star. It reflects the star's light but does not give out light of its own.

Poles. The ends of a magnet, or the places on the Earth, where the magnetic field is strongest.

Pollution. Harmful fumes, waste chemicals and garbage that dirty the environment.

Polymer. A material that is made up of very long molecules.

Pressure. The force acting on a certain area.

Pressurized container. A container used for holding liquids or gases at high pressure.

Protons. Particles inside the nucleus of an atom that carry a positive charge.

Radar. A device that measures the distance to an object, and the direction it is moving in, by bouncing radio waves off it.

Reaction. The equal and opposite force to any action.

Recycle. To use something again, rather than throwing it away, in order to conserve resources and reduce pollution.

Reflection. The way light or sound bounces.

Refraction. The way light rays bend when they pass through different materials.

Repel. To push something away.

Resistance, electrical. The way a material slows down electric current flowing through it.

Solar cell. A device that changes the energy from sunlight into electrical energy.

Solar system. The Sun and all the objects that orbit it, such as the planets.

Solution. A solid, liquid or a gas which is mixed up, or dissolved, in a liquid. The thing that is dissolved is called the solute, and the thing that does the dissolving is called the solvent.

Sonar. A system that uses the echoes of ultrasound waves to detect things that are underwater.

Spectrum. All the colors which together make up white light.

Speed. How far something goes in a certain time.

Star. An object in Space that gives out light of its own.

Static electricity. Electricity that stays in one place because it builds up on insulators.

Supersonic. Faster than the speed of sound.

Technology. The development of new things as a result of advances in science.

Telescope. A device which uses lenses to magnify distant objects.

Temperature. The measure of how hot or cold something is.

Thermal. Means 'to do with heat'. A thermal is a current of air that rises because it has been heated.

Thermal expansion. The way things get slightly larger when they are heated.

Thermometer. A device for measuring temperature.

Translucent. A translucent object lets some light pass through it.

Transparent. A transparent object lets all light pass through it.

Upthrust. The force pushing on an object that is in a liquid or gas. Things float because the upthrust pushes them up.

Vacuum. A completely empty space, with no solids, liquids or gases inside it.

Velocity. The speed at which an object is moving in a particular direction.

Vibration. A continuous, very fast, backwards and forwards movement.

Viscosity. The measure of how thick a liquid is.

Visible light. All the light and colors that can be seen by human eyes.

Volume. The amount of space that something takes up.

Zoology. The scientific study of animals.

Index

Answers

Page 5
Roman numbers

You may find Roman numerals used in the following places:
– to show the hours on clock faces
– to number the chapters in books
– to show a date on some coins
– to show dates on some monuments
– in the names of some kings and queens, for example, Louis XIV.

Page 7
How tall are you?

Their answers will vary depending on how large they are. People's arms, hands and feet are different sizes, so a unit of measurement based on one person's body is different from a unit of measurement based on someone else's body.

Page 11
Energy quiz

Many things in the picture have kinetic energy:
– the moving cars and truck
– the moving bicycle
– the ball thrown by the boy
– the sailing boats
– the boy on the swing
– the flying birds
– the falling raindrops

– the hair-drier uses electrical energy
– the lawn-mower uses electrical energy

– the lamp over the door gives out light energy
– the car headlights give out light energy

– the apples on the tree have potential energy
– the chicks in the nest have potential energy

Page 13
Find the energy changes

The wind's kinetic energy changes into the kinetic energy of the moving sailing boat.

The potential energy of the man on the jetty changes into kinetic energy as he dives into the water.

The chemical energy of fuel in the engine changes into the kinetic energy of the moving motor-boat.

Page 33
Quiz

On Earth, a mass of 1kg weighs about 10N. To calculate your weight in newtons, multiply your mass in kilograms by 10. For example, if you have a mass of 50kg, you would weigh 500N. (To be more exact, 1kg weighs 9.8N on Earth, so to calculate your weight exactly in newtons, multiply your mass in kilograms by 9.8.)

On the Moon, a mass of 1kg only weighs 1.6N. To calculate your weight on the Moon, multiply your mass in kilograms by 1.6. If you have a mass of 50kg, you would weigh 80N.

Page 40
Pressure

A sharp knife cuts into things better than a blunt knife because the push of the cutting edge is spread over a smaller area. This means that a sharp knife puts more pressure on an object than a blunt knife.

In the same way, nails have sharp points so that they can be easily hammered into things. The sharper the point of a nail, the higher the pressure, because it pushes on a smaller area.

Page 55
Picture trick

If you check with a ruler, you will find that both lines are exactly the same length.

First published in 1988 by Usborne Publishing Ltd., 83-85 Saffron Hill, London EC1N 8RT, England
Copyright © 1988 Usborne Publishing Ltd. American edition 1988.